T0222599

Springer Geography

The Springer Geography series seeks to publish a broad portfolio of scientific books, aiming at researchers, students, and everyone interested in geographical research.

The series includes peer-reviewed monographs, edited volumes, textbooks, and conference proceedings. It covers the major topics in geography and geographical sciences including, but not limited to; Economic Geography, Landscape and Urban Planning, Urban Geography, Physical Geography and Environmental Geography.

Springer Geography—now indexed in Scopus

More information about this series at https://link.springer.com/bookseries/10180

Courage Kamusoko

Optical and SAR Remote Sensing of Urban Areas

A Practical Guide

 Springer

Courage Kamusoko
Machida, Tokyo, Japan

ISSN 2194-315X   ISSN 2194-3168   (electronic)
Springer Geography
ISBN 978-981-16-5151-9   ISBN 978-981-16-5149-6   (eBook)
https://doi.org/10.1007/978-981-16-5149-6

This Springer imprint is published by the registered company Springer Nature Singapore Pte Ltd.
The registered company address is: 152 Beach Road, #21-01/04 Gateway East, Singapore 189721, Singapore

To my mother, who taught me that consistency is the key to progress!!

Preface

Cost-effective urban land cover mapping methods are required to produce up-to-date and accurate geospatial information for the United Nations Sustainable Development Goals (SDGs) and the United Nations New Urban Agenda (NUA). The increasing volume of Earth Observing (EO) data provides opportunities to map land cover in urban areas. However, conventional image processing and analysis using only desktop computers is quite difficult given the huge volumes of EO data. Therefore, open source software applications that include both cloud and desk computing are required in order to perform cost-effective remotely sensed image processing and classification.

The workbook is designed as a practitioner's guide and a critical resource for students, researchers, and others who are interested in applying optical and SAR data in order to improve land cover mapping in urban areas. Therefore, the workbook is intended to be hands-on, where users can explore data and methods to improve land cover mapping in urban areas. Although there are many freely available EO data, I focus on land cover mapping using Sentinel-1 C-band SAR and Sentinel-2 data from the Copernicus program under the European Space Agency (ESA). All remotely sensed image processing and classification procedures are based on open source software applications such QGIS and R as well as cloud-based platforms such as Google Earth Engine (GEE). Previous experience with open source software applications and GEE is recommended but not necessary. However, experience with GIS and remote sensing as well as basic statistical analysis is required to complete the laboratory exercises.

How is this Workbook Organized?

This workbook is organized into six chapters. Chapter 1 introduces geospatial machine learning, which is subdivided into five sections. Section 1.1 presents a brief introduction on geospatial machine learning, while Sect. 1.2 describes the study area and data. Section 1.3 provides preliminary image processing in Google Earth Engine (GEE). Finally, Sects. 1.4 and 1.5 provide the summary and additional exercises.

Chapter 2 covers exploratory image analysis and transformation. Section 2.1 provides a brief background on exploratory image analysis and image transformation. Next, Sect. 2.2 provides laboratory exercises in QGIS and R (i.e., preparing training data, creating spectral plots, computing spectral, and texture indices). Finally, Sects. 2.3 and 2.4 provide the summary and additional exercises.

Chapter 3 focuses on mapping urban land cover using multi-seasonal Sentinel-2 imagery. Section 3.1 provides a brief background. Next, Sect. 3.2 provides laboratory exercises 1 and 2, which are done in R. The final Sects. 3.3 and 3.4 provide the summary and additional exercises.

Chapter 4 focuses on mapping urban land cover using multi-seasonal Sentinel-1 imagery. Section 4.1 provides a brief background, while Sect. 4.2 provides laboratory exercises 1 and 2. Finally, Sects. 4.3 and 4.4 provide the summary and additional exercises.

Chapter 5 focuses on mapping urban land cover using multi-seasonal Sentinel-1 and Sentinel-2 imagery as well as other derived data such as spectral and texture indices. Section 5.1 provides a brief background, while Sect. 5.2 provides laboratory exercises 1 and 2. Finally, Sects. 5.3 and 5.4 provide the summary and additional exercises.

The final chapter 6 focuses on land cover classification accuracy assessment. Section 6.1 presents an overview on land cover classification accuracy assessment. Next, Sect. 6.2 provides accuracy assessment exercises. Finally, Sects. 6.3 and 6.4 provide the summary and additional exercises. An attempt has been made to organize the workbook in a general sequence of topics. Therefore, I encourage you to read the workbook in sequence from Chapter 1.

Conventions Used in this Workbook

R and Java commands or scripts are written in *Ubuntu Mono* font size 10 in italics, while the output is written in Ubuntu Mono font size 10. Note that long output from R and Java code is omitted from the workbook to save space. In some cases, I use small font sizes in Ubuntu Mono to show how the output or results would appear. This is just for illustration purposes. Readers will of course see the whole output when they execute the commands. The hash sign (# or //) at the start of a line of code indicates that it is a comment. Finally, all explanations are written in Liberation Serif font size 12.

Data and Online Resources

Boundary and training area data (shapefiles) used in this workbook are available. Furthermore, I provide additional online resources on R, QGIS, and GEE or remote sensing in the appendix.

Machida, Japan   Courage Kamusoko

Contents

Geospatial Machine Learning in Urban Environments: Challenges and Prospects

1

Abstract

Accurate and current land cover information is required to develop strategies for sustainable development and to improve quality of life in urban areas. The past decades has seen an increased availability of earth observation satellite (EOS) sensors (e.g., Sentinel-1 and Sentinel-2) as well as machine learning (ML) techniques (support vector machines, random forests) for land cover mapping. While significant progress has made to improve land cover mapping in urban areas, challenges still remain. The purpose of this chapter is to discuss briefly about geospatial machine learning in urban environments as well as some of its major challenges and prospects. The chapter will cover an introduction to geospatial ML (remote sensing image pre-processing and ML techniques), study area and data sets, hands-on exercises, summary, and additional exercises.

Keywords

Earth observation satellite sensors • Sentinel-1 • Sentinel-2 • Machine learning • Land cover • Urban areas

1.1 Introduction

1.1.1 Background

According to the United Nations, global urban population increased from 751 million in 1950 to 4.2 billion in 2018 (United Nations 2018). In the next decade, the global urban population is expected to increase to 60% (United Nations 2020; World Health Organization 2020). About 90% of urban growth will occur in less developed regions, such as East Asia, South Asia, and sub-Saharan Africa (UNHabitat 2020). To date, rapid urbanization has resulted in the increase of informal settlements and unplanned urban sprawl, especially in sub-Saharan Africa (Seto et al. 2011; United Nations 2018; UNHabitat 2020). As a result, most local government authorities fail to provide adequate infrastructure (transport and health facilities) and basic services such as clean water and sanitation (UNHabitat 2008). Furthermore, citizens living in densely populated and informal settlements are vulnerable to the outbreak of epidemics and global pandemics such as COVID-19 (Zerbo et al. 2020). Given the unplanned nature of informal settlements, overcrowding makes it difficult to maintain high hygiene standards as well as follow recommended measures such as social distancing and self-isolation (World Health Organization 2020; UNHabitat 2020). Therefore, practical policies and actions are needed to make urban environments inclusive, safe, and resilient as well as improve the quality of life urban dwellers. This requires accurate and timely geospatial information, which is critical for

Electronic supplementary material

The online version of this chapter (https://doi.org/10.1007/978-981-16-5149-6_1) contains supplementary material, which is available to authorized users.

planning and implementing sustainable urban development in light of the 2030 Agenda for Sustainable Development (United Nations 2019).

Geospatial or mapping agencies in less developed countries recognize the importance of geospatial information for sustainable urban planning and development. However, efforts to produce new or update geospatial information (e.g., large-scale topographic maps and land cover maps) have been constrained by poor funding as well as the high cost of acquiring data using conventional land surveys and aerial photography (Conitz 2000). It is noteworthy that most government policy makers do not prioritize investing in geospatial technology and information despite its contribution to spatial urban planning in particular, and sustainable development in general (UNHabitat 2008). Consequently, geospatial or mapping agencies in the less developed countries fail to produce timely, reliable, and accurate geospatial information. Furthermore, official development assistance (ODA) funding for major mapping projects has declined over the past decades (Kamusoko et al. 2021).

Recently, urban land cover mapping at a regional scale has increased in some less developed regions (Seto et al. 2011). This is because medium-resolution satellite remotely sensed data such as the Landsat series and Sentinel-1 and Sentinel-2 have relatively good global coverage and are available free of charge. Furthermore, advancement in machine learning methods such as random forests (Rodriguez-Galiano et al. 2012), support vector machines (Nemmour and Chibani 2006; Pal and Mather 2003) and deep learning (Yu et al. 2017; Ma et al. 2019) has also increased urban land cover mapping applications. However, most of the urban land cover mapping studies in the less developed regions have been done in capital cities (Mundia and Aniya 2005; Gamanya et al. 2009; Griffiths et al. 2010; Forkuor and Cofie 2011). This means that most cities or urban centers are still poorly quantified because mapping urban land cover at a national scale still remains difficult despite the availability of free satellite imagery and advanced ML techniques (Goldblatt et al. 2018). For example, high cost of collecting reliable reference data sets (training or validation) from field surveys and high spatial resolution imagery inhibits land cover mapping at a national scale. In addition, spectral confusion and mixed pixel problems still persist given the heterogeneous nature of the urban landscapes and the fragmented spatial configuration of small cities (Stefanov et al. 2001; Xian and Crane 2005). In most African urban cities, spectral confusion is a major problem because gravel (dirt) roads in informal settlements have similar spectral responses to those of bare vacant plots and croplands (Kamusoko et al. 2013). Nonetheless, recent technological advancement such as the availability of very high-resolution satellite imagery on Google Earth, cloud computing (e.g., Google Earth Engine) as well as the availability of microwave satellite imagery (Sentinel-1) are encouraging. This can be used to improve land cover mapping in urban areas.

1.1.2 Cloud and Desktop Open Source Software Applications

Earth observing satellites are increasing due to investments in large satellite missions and development of small satellite missions such as CubeSat. As a result, there is an increase in the volume of Earth Observing (EO) data that can be used to develop environmental applications (e.g., mapping and monitoring urban areas). However, conventional image processing and analysis using desktop computers is quite difficult given the huge volume of EO data. Therefore, open source software applications that include both cloud and desk computing are required in order to perform cost-effective remotely sensed image processing and classification. In this workbook, remotely sensed image processing and classification procedures are based on open source software applications such as QGIS and R as well as on Google Earth Engine (GEE) platform. QGIS is an open source desktop GIS application that provides data viewing, editing, and analysis capabilities. R is a free software programming language and software environment for statistical computing and graphics (R Development Core Team 2005). GEE is a cloud-based platform for planetary-scale geospatial analysis with massive computational capabilities (Gorelick et al. 2017).

In this workbook, we are going to focus on image processing and geospatial machine techniques. In this context, geospatial machine learning refers to the use of remotely sensed imagery and machine learning classifiers to map land cover in urban areas. This will also include a detailed accuracy assessment based on recommendations by Olofsson et al. (2014). The idea is to provide a hands-on approach in order to improve urban land cover mapping based on multi-seasonal Sentinel-1 and Sentinel-2 imagery, derived variables (spectral and texture indices), and machine learning classifiers such as random forests. An attempt is made to focus on common problems related to land cover classification such as spectral confusion, machine learning model performance, and accuracy assessment.

1.2 Study Area and Data

1.2.1 Harare Metropolitan Area

According to the Zimbabwe National Statistical Agency (ZimStats 2012), "urban" refers to a designated urban areas with a compact settlement patterns with more than 2500 inhabitants of which 50% are employed in non-agricultural sector. Zimbabwe's urban landscape encompasses the large metropolitan areas of Harare and Bulawayo, cities or municipalities and towns. In addition, about 472 small urban centers are available in the form of "growth points," district service centers and rural service centers (Infrastructure and Cities for Economic Development 2017; Mbiba 2017).

Harare metropolitan area (Fig. 1.1) comprises the City of Harare—which is the capital and largest city in Zimbabwe, Chitungwiza municipality, Ruwa and Epworth Local Boards. The metropolitan area is characterized by a warm, rainy season from November to March; a cool, dry season from April to August; and a hot, dry season in October. Daily temperatures range from about 7 to 20 °C in July (coldest month), and from 13 to 28 °C in October (hottest month). The metropolitan area receives a mean annual rainfall ranging from 470 to 1350 mm between November and March. The population in the City of Harare increased from approximately 310,360 in 1962 to 1,435,784 in 2012, while the population of Chitungwiza grew from approximately 14,970 in 1969 to 354,472 in 2012 (Patel 1998). The population of Epworth increased from 114,067 in 2002 to 161,840 in 2012 (ZimStats 2012). Harare Metropolitan Area represents over 47% of total urban population in Zimbabwe. The City of Harare has many major primary and secondary industries, while Chitungwiza has a small industrial park.

1.2.2 Satellite Imagery

Sentinel-1 and Sentinel-2 imagery are derived from a constellation of satellites developed by the European Space Agency (ESA) under the Copernicus program (ESA 2015).

1.2.2.1 Sentinel-1 Data

Sentinel-1 mission comprises a constellation of Sentinel-1A and Sentinel-1B satellites, which provides 12-day (ground track) repeat cycle for one satellite, and 6-day (ground track) repeat for two satellites (ESA 2020). In principle, Sentinel-1 provides imagery in both ascending (south to north) and descending (north to south) orbits, especially over Europe (ESA 2020). In our study area, a few descending orbit acquisitions are available. The Sentinel-1 constellation provides C-band (5.6 cm) synthetic aperture radar (SAR) (ESA 2015). Sentinel-1 data is acquired in the following four modes:

Fig. 1.1 Location of Harare Metropolitan area (central image shows elevation): **a** rainy season S-1; **b** post-rainy season S-1 **c** dry season S-1; **d** rainy season S-2; **e** post-rainy season S-2; and **f** dry season S-2 data. Note that S-1 imagery is displayed in false color VV (red), VH (green), and VV (blue) for visualization purposes only

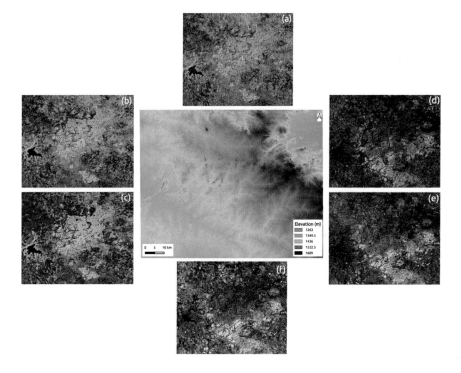

- Strip Map (SM): 80 km swath, 5 × 5 m spatial resolution;
- Interferometric Wide Swath (IW): 250 km swath, 5 × 20 m spatial resolution;
- Extra-Wide Swath (EW): 400 km swath, 20 × 40 m spatial resolution; and
- Wave (WV): 20 × 20 km, 5 × 5 m spatial resolution.

Sentinel-1 data products are available in dual polarization (VV + VH or HH + HV) and single polarization (VV or HH) for SM, IW, and EW modes (ESA 2020). Polarization refers to orientation or direction of the electric component in an electromagnetic wave (Sabins 1997; Jensen 2000). In most cases, radar signals are transmitted and received at vertical (V) and/or horizontal (H) polarization. In general, VV and HH are known as co-polarization (or like-polarization) backscatter components, while VH and HV are cross-polarization backscatter components. Single co-polarization VV refers to vertical transmit and vertical receive, while HH refers to horizontal transmit and horizontal receive (Eckardt et al. 2019). In this workbook, we are going to use Level-1 Ground Range Detected (GRD) data acquired in IW mode only.

1.2.2.2 Sentinel-2 Data

Sentinel-2 is a wide swath, high-resolution, multispectral imaging mission with a global 5-day revisit frequency (ESA 2015). The Sentinel-2 Multispectral Instrument (MSI) provides 13 spectral bands (ESA 2015). That is four bands at 10 m, six bands at 20 m, and three bands at 60 m spatial resolution (Table 1.1).

1.3 Preliminary Image Processing in GEE

The following labs will focus on preliminary image processing in GGE. This involves preparing Sentinel-2 and Sentinel-1 image collections in GEE. In order to use the GEE platform, you need to be registered for a GEE account. Following is a brief description of the registration steps.

First, go to the **GEE sign up page** and enter the email address you want to use for your GEE account. Note that a Gmail account is recommended. Second, enter your email, your affiliation, and country/region. Third, review the terms, verify that you are not a robot, and click "Submit." Finally, check your email, including your spam folder, for a link from the Google Developer's Team. The confirmation email will have directions on how to access the **Code Editor**.

Generally, users can run geospatial analysis on the GEE platform through the **Code Editor** or **Explorer**. The **Code Editor** is a web-based integrated development environment (IDE) for writing and running scripts, while the **Explorer** is a lightweight web application for exploring Earth engine data catalog and performing simple analyses. In this workbook, we are going to use JavaScript (not Java, which is a programming language used for web development) in the **Code Editor**. Users can read more about the **Code Editor** in the **Earth Engine's Developer Guide**, which offers a comprehensive guide.

Table 1.1 Spectral bands for the Sentinel-2 sensors

Sentinel-2 bands	Sentinel-2A	Sentinel-2B	
	Central wavelength (nm)	Central wavelength (nm)	Spatial resolution (m)
Band 1—Coastal aerosol	442.7	442.2	60
Band 2—Blue	492.4	492.1	10
Band 3—Green	559.8	559.0	10
Band 4—Red	664.6	664.9	10
Band 5—Vegetation red edge	704.1	703.8	20
Band 6—Vegetation red edge	740.5	739.1	20
Band 7—Vegetation red edge	782.8	779.7	20
Band 8—NIR	832.8	832.9	10
Band 8A—Narrow NIR	864.7	864.0	20
Band 9—Water vapor	945.1	943.2	60
Band 10—SWIR—Cirrus	1373.5	1376.9	60
Band 11—SWIR	1613.7	1610.4	20
Band 12—SWIR	2202.4	2185.7	20

While it is not necessary to formally learn JavaScript to perform image analysis in GEE, there are many online tutorials that teach JavaScript if you want to advance your knowledge.

1.3.1 Lab 1. Processing Sentinel-2 Imagery

In this workbook, we are going to use multi-seasonal Sentinel-2 imagery collection scenes acquired between January and October 2020. The aim of the lab is to create quarterly multi-seasonal composite imagery (i.e., 3-monthly composites for the rainy, post-rainy, and dry season). On completing the exercise, you should be able to load and visualize Sentinel-2 Imagery in Google Earth Engine (GEE) as well as to export the imagery so that it can be visualized in R or GIS software.

Objectives

- Learn how to access Sentinel-2 images,
- Display Sentinel-2 images, and
- Export Sentinel-2 images for use in other GIS software.

Procedure

In order to access the GEE environment, type the address "https://code.earthengine.google.com" in your Chrome browser (Fig. 1.2).

Next, access the Earth Engine Code Editor (Fig. 1.3) using your user account. Notice that the Earth Engine Code Editor environment is divided up into four panels: (1) panel 1 comprises tabs for **Scripts**, **Docs,** and **Assets**; (2) panel 2 is for writing and running Javascript commands; (3) panel 3 consist of the Console, Inspector, and Tasks tabs; and (4) panel 4 comprises the map interface.

Fig. 1.2 Access to the Earth Engine Code Editor environment

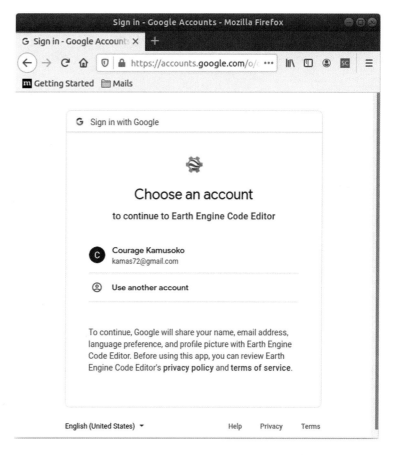

Fig. 1.3 Earth Engine Code Editor environment

Fig. 1.4 Import study area boundary

First, upload the study area boundary into the Earth Engine Assets. The boundary will be used to clip the imagery collection.

Click *Assets > New > Shapefiles (.shp,.shx,.dbf,.prj or.zip)*. Then, select the boundary shapefile. Note the allowed extensions are shp, zip, dbf, prj, shx, cpg, fix, qix, sbn, or shp.xml (Fig. 1.4).

Next, search for "Harare" in the search bar (Fig. 1.5). A map showing Harare appears.

Next, click the **Add a Marker** icon 🎯 and then click the center of Harare. A point is added in the center of Harare (Fig. 1.6). Make sure that you click on **Exit** on the drawing tool when you are done.

Notice that the "geometry" comes by default and coordinates appear (Fig. 1.6).

```
var geometry: Point (31.05; -17.82)
```

Go to **Assets** and then click on "Hre_Boundary" to import the study area boundary. Make sure that you change the name from "table" to "boundary" (Fig. 1.7).

Fig. 1.5 Search for study area (e.g., Harare, Zimbabwe)

Fig. 1.6 Marking a point in the center of the study area

```
var boundary: Table Users/Kamas72_ML_Zim_Cities/Hre_Boundary
```

Next, load a stack of images called **Image Collections**. Each data source available on GEE has its own **Image Collection** and **ID** (e.g., *COPERNICUS/S2_SR*). Start by typing the following lines of code (Fig. 1.8). I recommend typing instead of just copying and pasting. This helps to learn writing scripts.

```
// Load Sentinel-2 surface reflectance data.
var s2 = ee.ImageCollection('COPERNICUS/S2_SR');
```

Fig. 1.7 Import the study area boundary

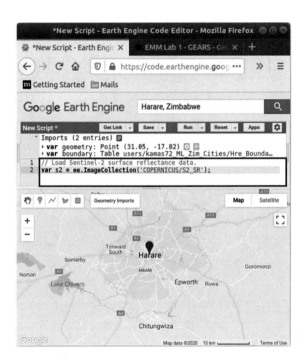

Fig. 1.8 Loading Sentinel-2 imagery

In the JavaScript, two backslashes (//) indicate comments that will be ignored when running the code. These comments are included in the code in order to inform what the code will do. This is important when you share your code.

Define a function that can be used to mask clouds using the Sentinel-2 QA band. The "*maskS2clouds*" function can be applied to each image in the **imageCollection**. Note that functions need to explicitly **return** the final output.

```
// Function to mask clouds using the Sentinel-2 QA band.
function maskS2clouds(image) {
  var qa = image.select('QA60');
  // Bits 10 and 11 are clouds and cirrus, respectively.
```

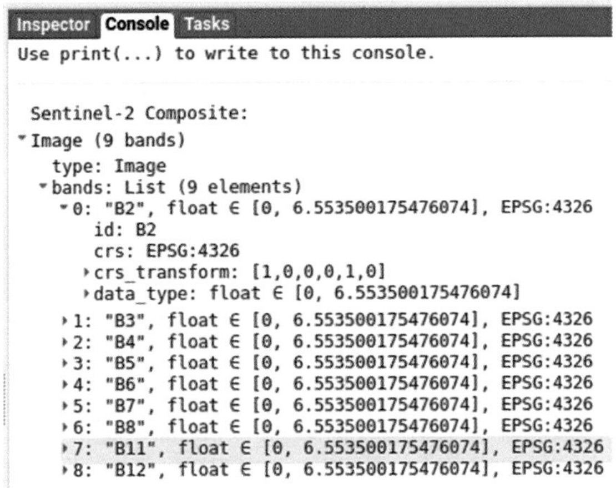

Fig. 1.9 Sentinel-2 image propertied displayed in the console

```
var cloudBitMask = ee.Number(2).pow(10).int();
var cirrusBitMask = ee.Number(2).pow(11).int();

// Both flags should be set to zero, indicating clear conditions.
var mask = qa.bitwiseAnd(cloudBitMask).eq(0).and(
        qa.bitwiseAnd(cirrusBitMask).eq(0));

// Return the masked and scaled data.
return image.updateMask(mask).divide(10000);
}
```

The QA60 is a bitmask band with cloud mask information (bits 10 and 11). Bit 10 represents "Opaque clouds," where 0 means "No opaque clouds," and 1 means "Opaque clouds present." Bit 11 represents "Cirrus clouds," where 0 means "No cirrus clouds" and 1 means "Cirrus clouds present."

Filter the **ImageCollection** by date, sort by a metadata property called "CLOUDY_PIXEL_PERCENTAGE", and use the mask that you created above. You can also select the bands that you will need to use (see Table 1.1).

```
// Map the cloud masking function over one year of data
var S2 = s2.filterDate('2020-01-01', '2020-03-30')
            .filter(ee.Filter.lt('CLOUDY_PIXEL_PERCENTAGE', 10))
            .map(maskS2clouds)
            .select('B2', 'B3', 'B4', 'B5', 'B6', 'B7', 'B8', 'B11', 'B12');
```

Create a median composite image and print the scene metadata to the **Console**.

```
// Create median composite image
var composite = S2.median();
// Print the image to the console
print('Sentinel-2 Composite:', composite);
```

Next, activate the **Console.** Then, click the small triangle icon (▶) next to the image name in order to see more information stored in a composite image object. Expand the bands and inspect the long list of metadata items stored as properties of the composite image (Fig. 1.9).

Display the Sentinel-2 median image in false color (Fig. 1.10). In the code below, *bands* represent a list of three bands to display as red (band 8), green (band 4), and blue (band 3), respectively. Note that band 8 is the near infrared (NIR), band 4 is red, and band 3 is green (Table 1.1). We can also clip the image using the study area boundary.

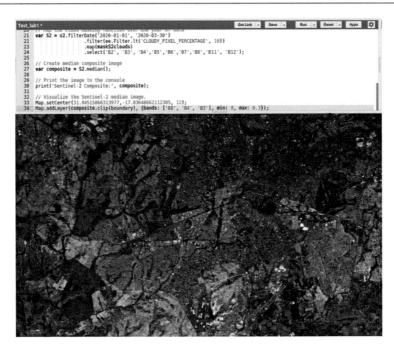

Fig. 1.10 Sentinel-2 median image displayed in false color

```
// Visualize the Sentinel-2 median image.
Map.setCenter(31.04515866313977,-17.83648662112305, 12);
Map.addLayer(composite.clip(boundary), {bands: ['B8', 'B4', 'B3'], min: 0, max: 0.3});
```

You can also play around with the "**Inspector**" tool (Fig. 1.11), which allows you to query map layers at a point. Click on the "**Inspector**" tab in the upper right panel to activate it. Then, click anywhere within the **Map Viewer**. The coordinates of your click will be displayed, along with the value for the map layers at that point.

```
Inspector  Console  Tasks
▸Point (31.0989, -17.8185) at 38m/px
▾Pixels
  ▾Layer 1: Image (9 bands) 🔗
      B2: 0.023800000548362732
      B3: 0.048000000041723251
      B4: 0.023199999970495701
      B5: 0.071999996900055847
      B6: 0.28279998898506165
      B7: 0.3702999949455261
      B8: 0.4063999508857727
      B11: 0.17710000276565552
      B12: 0.08789999783039093
▾Objects
  ▾Layer 1: Image (9 bands)
      type: Image
     ▾bands: List (9 elements)
       ▸0: "B2", float ∈ [0, 6.553500175476074], EPSG:4326
       ▸1: "B3", float ∈ [0, 6.553500175476074], EPSG:4326
       ▸2: "B4", float ∈ [0, 6.553500175476074], EPSG:4326
       ▸3: "B5", float ∈ [0, 6.553500175476074], EPSG:4326
       ▸4: "B6", float ∈ [0, 6.553500175476074], EPSG:4326
       ▸5: "B7", float ∈ [0, 6.553500175476074], EPSG:4326
       ▸6: "B8", float ∈ [0, 6.553500175476074], EPSG:4326
       ▸7: "B11", float ∈ [0, 6.553500175476074], EPSG:4326
       ▸8: "B12", float ∈ [0, 6.553500175476074], EPSG:4326
     ▾properties: Object (1 property)
         system:index: 0
```

Fig. 1.11 Querying Sentinel-2 median image using the **Inspector** tool

Fig. 1.12 Exporting from the **Tasks** tab

Next, export the composite image (GeoTIFF) so that it can be displayed in GIS software or other software. You can export the composite image GEE Asset folder for use within GEE or to your personal Google Drive or Google Cloud Storage accounts. All exports are sent to the '**Tasks**' tab in the upper right panel (Fig. 1.12).

```
// Export the image, specifying scale and region.
Export.image.toDrive({
  image: composite,
  description: 'S2_Jan_Mar2020',
  scale: 10,
  crs: "EPSG:32736", // EPSG:32736 - WGS 84 / UTM zone 36 - Projected
  maxPixels: 1066537920,
  region: boundary
});
```

Notes: Run individual exports from the 'Tasks' tab to avoid overwhelming the system. You can also change filenames and other parameters. When exporting to Google Drive, GEE will find the named folder specified and does not need the full file path. If this folder does not yet exist, it will create it for you in your Drive. In addition, you can define the maxPixels parameter. For example, you can increase **maxPixels** if the default value is too low for the output image.

Finally, save your script by clicking "**Save**". The script will be saved in your private repository (Fig. 1.13). This script will be available and accessible the next time you log in to GEE.

1.3.2 Lab 2. Exploring Sentinel-1 Imagery

The purpose of this lab is to explore Sentinel-1 image collection. Lab 2 consist of four exercises. Lab 2a focuses on accessing Sentinel-1 imagery available in the image collection as well as selecting Sentinel-1 sensor orbits. Lab 2b examines in detail Sentinel-1 ascending and descending orbits. Lab 2c focuses on polarization, while lab 2d deals with exporting Sentinel-1 imagery. On completing the exercise, you should be able to load and visualize Sentinel-1 imagery in GEE as well as export the imagery so that it can be used in GIS software.

Fig. 1.13 Save script in your repository

1.3.3 Lab 2(a). Accessing Sentinel-1 Imagery

Objectives

- Learn how to access Sentinel-1 images

Procedure

Start by searching "Harare" in the search bar and adding the map center for the study area (Fig. 1.14) as we have done in lab exercise 1.

```
// Add map center for the project area.
Map.centerObject(Harare,12);
```

Go to **Assets** and then click on 'Hre_Boundary' to import the study area boundary. Make sure that you change the name from "table" to "boundary." Remember, this what we did in lab exercise 1.

Next, load a collection of Sentinel-1 images (COPERNICUS/S1_GRD) and then filter the **ImageCollection** by boundary. This will create an object **collection_S1** that contains the Sentinel-1 image collection. You can use the ***print*** command to check list of Sentinel-1 images in the **collection_S1** object.

```
// Load the Sentinel-1 Image Collection and filter to the geometry (center point).
var collection_S1 = ee.ImageCollection('COPERNICUS/S1_GRD').filterBounds(Harare);
print('S1 images Harare',collection_S1);
```

Next, activate the **Console** to check the metadata for **ImageCollection** and examine the information you printed (Fig. 1.15). Using the drop down arrows you can see that **ImageCollection** contains 136 elements (which is the Sentinel-1 image collection).

Next, expand **features** tab under the **ImageCollection**. A list with 136 elements of 'COPERNICUS/S1_GRD' is displayed (Fig. 1.16).

Fig. 1.14 Center of the study area

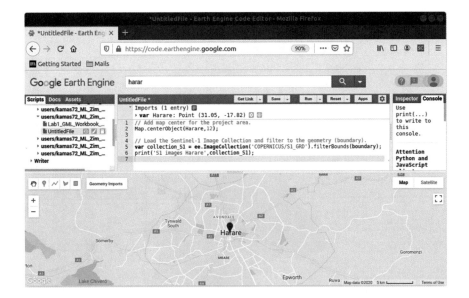

Fig. 1.15 Sentinel-1 image
collection and properties
displayed in the **Console**

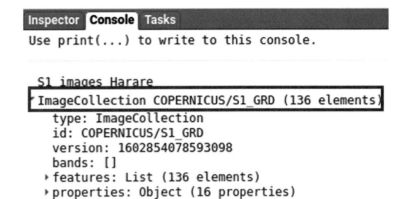

Fig. 1.16 Sentinel-1 image
collection and properties
displayed in the **Console**

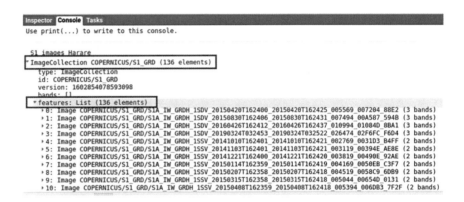

Note that the file name gives important information. Therefore, let us take a look at the first file name "*S1A_IW_GRDH_1SDV_20150420T162400_20150420T162425_…*". First, S1A denotes the Sentinel-1A instrument, while IW refers to the mode—that is, interferometric wide (IW) swath mode (ESA 2020). Second, GRDH refers to the product type and resolution class. In this case, we have the Ground Range Detected (GRD) product with a high (H)-resolution class. Therefore, the product consists of multilook intensity data that has been projected to ground range using the Earth ellipsoid model WGS84. Next, 1SDV refers to the processing level (1), the product class (i.e., S refers to SAR Standard), and DV refers to dual VV + VH polarization. The product start and stop date and times are shown as fourteen digits representing the date and time, separated by the character "T". For example, 20150420T162400 refers to the start date and time (April 20, 2015, at 16:24:00), while 20150420T162425 refers to the stop date and time (April 20, 2015, at 16:24:25). Last but not least, 005569, 007204, and 88E2 show the absolute orbit number, mission take ID, and product unique identifier (ESA 2020). The information in the **Console** (Fig. 1.16) shows us that Sentinel-1 data or features from 0 to 10 was acquired using Sentinel-1A instrument. In addition, all the Sentinel-1 data in our study area are in IW swath mode and have dual VV + VH and single polarization. Note that you can also click on **bands** and **properties**, which will give more additional information such orbit properties.

Next, filter the **ImageCollection** based on ascending and descending orbits. We will examine Sentinel-1 sensor orbits in lab 2b.

```
// Filter to get the imagery from different look angles (ascending mode).
var collection_S1_ASC = collection_S1.filter(ee.Filter.eq('orbitProperties_pass', 'ASCENDING'));
print('S1 IW images from ascending orbit',collection_S1_ASC);
```

```
// Filter to get the imagery from different look angles (descending mode).
var collection_S1_DSC = collection_S1.filter(ee.Filter.eq('orbitProperties_pass', 'DESCENDING'));
print('S1 IW images from descending orbit',collection_S1_DSC);
```

The Console shows that 135 Sentinel-1 images were collected in ascending orbit and only one in descending orbit (Fig. 1.17). The descending orbit imagery was acquired on March 24, 2019, (S1_GRD/S1A_IW_GRDH_1SDV_**20190324**T032453_20190324).

1.3.4 Lab 2(b). Examining Sentinel-1 Orbit Properties

Objectives

- Display Sentinel-1 imagery and
- Examine Sentinel-1 orbit properties.

Procedure

In this exercise, we are going explore the Sentinel-1 imagery in both ascending and descending orbits. In order to do that, we will select both ascending and descending orbit Sentinel-1 imagery within the same acquisition period. Remember in the previous exercise, we observed that one descending orbit Sentinel-1 imagery was acquired on April 24, 2019. Therefore, we are also going to define the same acquisition date for the ascending orbit.

Start by searching "Harare" in the search bar and adding the map center for the study area (Fig. 1.14).

We are going to specify one day time interval since we know that specific date of image acquisition.

```
// Ascending and descending acquisitions.
// Specify the time interval.
var start_date = ee.Date('2019-03-24');
var end_date = start_date.advance(1, 'days');
var date_filter = ee.Filter.date(start_date, end_date);
```

Next, filter the imagery according to the acquisition date and VV polarization.

```
// Filter Sentinel-1 collection according to specified criteria.
var collection_S1 = ee.ImageCollection('COPERNICUS/S1_GRD')
    .filter(date_filter)
    .filter(ee.Filter.listContains('transmitterReceiverPolarisation', 'VV'))
    .select(['VV','angle']);
```

Fig. 1.17 Sentinel-1 image collection in ascending and descending orbits

```
Inspector  Console  Tasks
Use print(...) to write to this console.

S1 images Harare                                               JSON
▶ ImageCollection COPERNICUS/S1_GRD (136 elements)            JSON

S1 IW images from ascending orbit                             JSON
▶ ImageCollection COPERNICUS/S1_GRD (135 elements)            JSON

S1 IW images from descending orbit                            JSON
▶ ImageCollection COPERNICUS/S1_GRD (1 element)               JSON
```

After that, we are going to specify parameters so that we can visualize the imagery as well as its angle in both ascending and descending orbits.

```
// Define the Sentinel-1 data layers.
var label1 = 'Ascending Orbit';
var label1_vv = 'Ascending Orbit - angle';
var collection1 = collection_S1
    .filter(ee.Filter.eq('orbitProperties_pass', 'ASCENDING'));
var vis_params1 = {bands:'VV', min:-25, max:5};
var vis_params1_angle = {bands:'angle', min:18.3, max:46.8};

var label2 = 'Descending Orbit';
var label2_vv = 'Descending Orbit - angle';
var collection2 = collection_S1
    .filter(ee.Filter.eq('orbitProperties_pass', 'DESCENDING'));
var vis_params2 = {bands:'VV', min:-25, max:5};
var vis_params2_angle = {bands:'angle', min:18.3, max:46.8};
```

Here, we are going to create image 1 (ascending orbit) and image 2 (descending orbit) objects.

```
// Create the map objects, link them, and display them.
var map1 = ui.Map().add(ui.Label(label1, {position:'middle-left'}));
map1.addLayer(collection1, vis_params1, label1, true);
map1.addLayer(collection1, vis_params1_angle, label1_vv, false);

var map2 = ui.Map().add(ui.Label(label2, {position:'middle-right'}));
map2.addLayer(collection2, vis_params2, label2, true);
map2.addLayer(collection2, vis_params2_angle, label2_vv, false);
var linker = ui.Map.Linker([map1, map2]);
var split_panel = ui.SplitPanel({
  firstPanel: map1,
  secondPanel: map2,
  wipe: true,
});
map1.setCenter(31.05,-17.8, 12); // Harare
```

Finally, we are going to display the imagery using the split window panel (Fig. 1.18). This will allow us to visualize the both images acquired in ascending and descending orbits.

```
// Add the split panel to the UI.
ui.root.widgets().reset([split_panel]);
```

Observations: Although the ascending and descending orbit Sentinel-1 images (Fig. 1.19) were acquired with the same polarization and during the same period or season, the difference in appearance is conspicuous. For example, the city center is brighter in the ascending orbit imagery. This is due to the sensor look direction in relation to the orientation of the features on the ground. Note that, the Sentinel-1 SAR sensor has a right-side looking geometry (ESA 2015). In ascending orbit (Fig. 1.20)—that is, the south to north movement—the Sentinel-1 antenna look direction is toward the west. Therefore, the city center has bright backscatter signatures caused by the tall buildings (due to double bounce) and street orientation (cardinal effect). This observation means interpreters of radar images in urban areas must be aware of the nature of these geometric relationships in their study area. We are going to discuss more about double bounce and the "cardinal effect" in lab 2c.

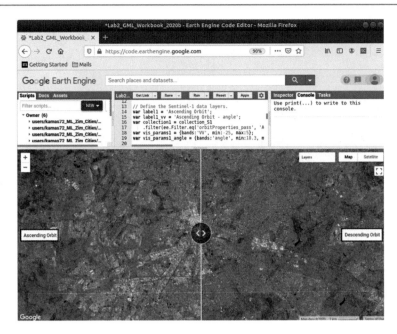

Fig. 1.18 Sentinel-1 imagery acquired in ascending and descending orbits, displayed in split window viewer

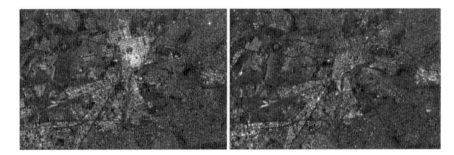

Fig. 1.19 Sentinel-1 VV imagery. The left side shows imagery acquired in ascending orbit, while the right side shows imagery acquired in descending orbit

You can also check the orbit angle using the *Inspector* tool (Fig. 1.21).

Observations: Figure 1.21 shows the radar image brightness expressed as backscatter coefficient ($\sigma°$ or sigma-nought) in decibels (dB). This represents target backscattering per unit ground area (the unit of $\sigma°$ is [m2/m2]). Note that the incidence angle shown in Fig. 1.21 is 36°. However, the incidence angle in the study area varies between 33° and 41°. Generally, returns are normally strong at low incidence angles and decrease with increasing incidence angle. That is higher (brighter) returns are in the near-range part of the image (closest to the satellite), while and lower (darker) are in the far-range of the image, which are further away from the satellite (CEOS 2018). It should be noted that SAR backscatter is influenced by the terrain structure and surface roughness. Rough surfaces usually scatter energy and return a significant amount back to the antenna resulting in a bright feature. However, flat surfaces reflect the signal away resulting in a dark feature.

Fig. 1.20 Sentinel-1 scenes in ascending (left) and descending (right) orbits. You can simply zoom out in order to see scene orbit path in GEE platform

Fig. 1.21 Sentinel-1 imagery properties

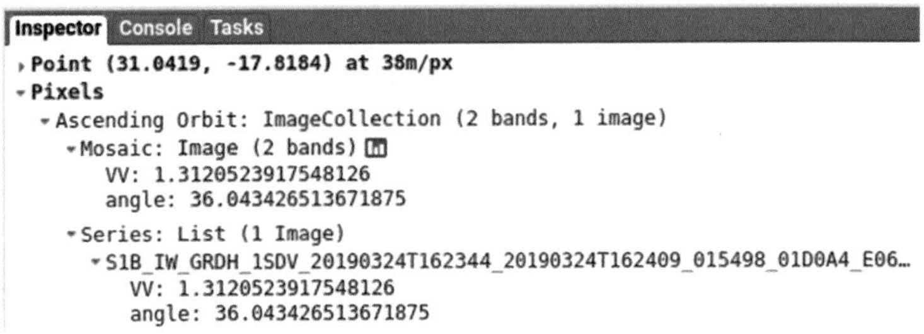

```
Inspector  Console  Tasks
▸ Point (31.0419, -17.8184) at 38m/px
▾ Pixels
   ▾ Ascending Orbit: ImageCollection (2 bands, 1 image)
      ▾ Mosaic: Image (2 bands) ⏏
          VV: 1.3120523917548126
          angle: 36.043426513671875
      ▾ Series: List (1 Image)
         ▾ S1B_IW_GRDH_1SDV_20190324T162344_20190324T162409_015498_01D0A4_E06…
             VV: 1.3120523917548126
             angle: 36.043426513671875
```

1.3.5 Lab 2(c). Examining Sentinel-1 Polarization

Objectives

- Examine Sentinel-1 polarization and
- Display the imagery in false color.

Procedure

In this lab, we are going to examine Sentinel-1 polarization. First, add 'Hre_Boundary' from **Assets** and change the name from "table" to "boundary."

Next, specify one day time interval since we know that specific data of image acquisition.

```
// Visualize different SAR polarization.
// Define the time interval.
var start_date = ee.Date('2020-01-01');
var end_date = start_date.advance(6, 'days');
var date_filter = ee.Filter.date(start_date, end_date);
```

Next, filter the imagery according to the acquisition data, study area boundary, the orbit pass, and polarization (VV and VH). Note that we have limited the imagery to study area only.

```
// Filter Sentinel-1 collection according to specified criteria.
var collection_S1 = ee.ImageCollection('COPERNICUS/S1_GRD')
      .filter(date_filter)
      .filterBounds(boundary)
      .filter(ee.Filter.eq('orbitProperties_pass', 'ASCENDING'))
      .select(['VV','VH']);
```

After that, we are going to specify parameters so that we can visualize the VV and VH imagery.

```
// Define the Sentinel-1 data layers.
var label1 = 'VH Polarization';
var collection1 = collection_S1
    .filter(ee.Filter.listContains('transmitterReceiverPolarisation', 'VH'));
var vis_params1 = {bands:'VH', min:-25, max:5};
print('VH Collection 1 imagery', collection1);

var label2 = 'VV Polarization';
var collection2 = collection_S1
    .filter(ee.Filter.listContains('transmitterReceiverPolarisation', 'VV'));
var vis_params2 = {bands:'VV', min:-25, max:5};
print('VV Collection 2 imagery', collection1);
```

We will also create image 1 (VH) and map 2 (VV) objects as we have done in the previous exercise (Lab 2b).

```
// Create, link and display the map objects.
var map1 = ui.Map().add(ui.Label(label1, {position:'middle-left'}));
map1.addLayer(collection1, vis_params1, label1);

var map2 = ui.Map().add(ui.Label(label2, {position:'middle-right'}));
map2.addLayer(collection2, vis_params2, label2);

var linker = ui.Map.Linker([map1, map2]);
var split_panel = ui.SplitPanel({
  firstPanel: map1,
  secondPanel: map2,
  wipe: true,
});
map1.setCenter(31.0479, -17.8263, 11); // Harare
```

Finally, we are going to display the imagery using the split window panel (Fig. 1.22). This will allow us to visualize both VH and VV images.

```
// Add the split window panel to the user interface.
ui.root.widgets().reset([split_panel]);
```

Observations: Generally, the VH imagery appears significantly darker than the VV imagery. Figure 1.22 shows that the city center is brighter in the VV imagery than in the VH imagery. This due to the specular double-bounce and the "cardinal" effects. Double bounce occurs when the radar signal is first reflected specularly as it hits smooth surfaces such as roads and pavements in cities (CEOS 2018). Then the specularly reflected signal bounces off the sides of buildings and is reflected to the SAR antenna (CEOS 2018). This causes most of the radar signal to return to the sensor, which results in high backscatter and bright areas in the SAR imagery (Sabins 1997, Jensen). The "cardinal" effect refers to the tendency of SAR sensor to produce a very strong backscatter from a city street pattern oriented orthogonal to the SAR antenna (Bryan 1979; Hardaway et al. 1982; Sabins 1997; Jensen 2000). In this study area, the Rainbow Towers hotel acts as corner reflector since it appears brighter in both VV and VH polarized images (Fig. 1.23a). However, the woodland area in Kopje area is not visible in both VV and VH polarized images (Fig. 1.23b). This is because smoother woodlands canopies (during the rainy season) usually display a lower backscatter at both VV and VH polarization. The Harare Garden Park has a medium return in the VV polarized image and thus visible, whereas it is not visible in VH polarized imagery (Fig. 1.23c). The original streets in city center were designed as rectangular grid aligned to the north–south (Christopher 1977). Consequently, the streets with this orientation appears brighter in the VV imagery due to a strong SAR backscatter (Fig. 1.23d).

Fig. 1.22 Sentinel-1 SAR
backscatter—VH polarization
(left); VV polarization (right)

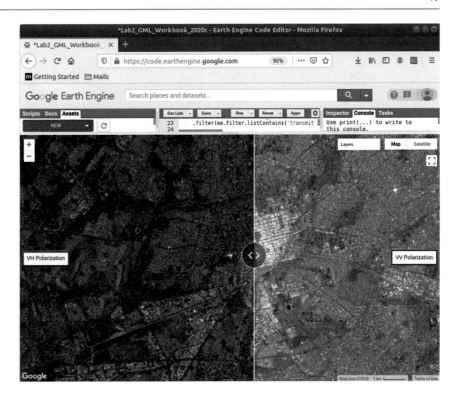

Fig. 1.23 Sentinel-1 image
displayed in VV and VH
polarization, Google Satellite and
Google Maps. Note **a** is the
Rainbow Towers hotel, **b** Kopje
area, **c** Harare Gardens Park, and
d city center

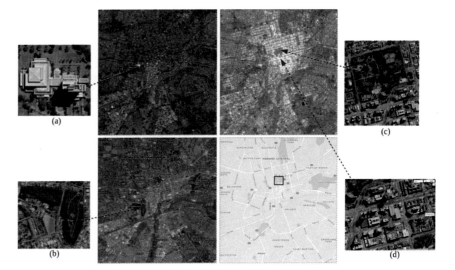

1.3.6 Lab 2(d). Exporting Sentinel-1 Imagery

In this workbook, we are also going to use multi-seasonal Sentinel-1 Ground Range Detected (GRD) scenes that were acquired between January and October 2020. Therefore, this lab will focus on creating single co-polarization (VV) and cross-polarized (VH) quarterly composite imagery (for the rainy, post-rainy and dry season). On completing the exercise, you should be able to display Sentinel-1 composite in false RGB as well as to export the imagery.

Objectives

- Display Sentinel-1 composite and
- Export Sentinel-1 images.

Procedure

First, add the map center for the study area (Fig. 1.24).

```
// Add map center for the project area.
Map.centerObject(Harare,12);
```

Go to **Assets** and then click on 'Hre_Boundary' to import the study area boundary.

Next, filter by ascending orbit pass and date. We are creating the rainy season Sentinel-1 image collection.

```
// Filter Sentinel-1 collection according to specified criteria.
var collection_S1 = ee.ImageCollection('COPERNICUS/S1_GRD')
    .filter(ee.Filter.date('2020-01-01', '2020-03-31'))
    .filterBounds(boundary)
    .filter(ee.Filter.eq('orbitProperties_pass', 'ASCENDING'))
    .select(['VV','VH']);
print('S1 images Harare',collection_S1)
```

Next, retrieve the latest SI imagery and display it.

```
//Get the latest image only
var collection_S1_first=ee.Image(collection_S1.sort('system:time_start',false)
  .first());
print('S1IWlatest acquired image', collection_S1_first);

//Display the latest images only
Map.addLayer(collection_S1_first, {'bands': 'VV', min: -15,max: 0}, 'Sentinel-1 VV');
Map.addLayer(collection_S1_first, {'bands': 'VH', min: -25,max: -5}, 'Sentinel-1 VH');
```

Then, display the Sentinel-1 RGB color composite using the "VV, VH, VV/VH" combination (Fig. 1.25).

```
//Create RGB composite that includes a VV/VH ratio band
var composite = collection_S1_first.addBands(collection_S1_first.select('VV')
  .subtract(collection_S1_first.select('VH')).select([0], ['Ratio']))
print('RGBComposite',composite);
```

Fig. 1.24 Location of the center of the study area

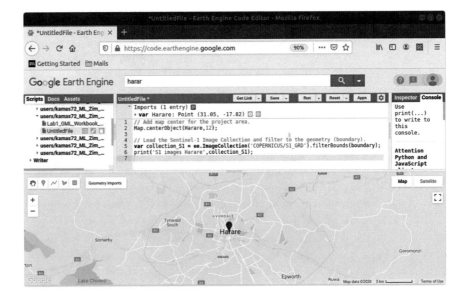

Fig. 1.25 Sentinel-1 VV, VH,
and VV/VH polarization
composite

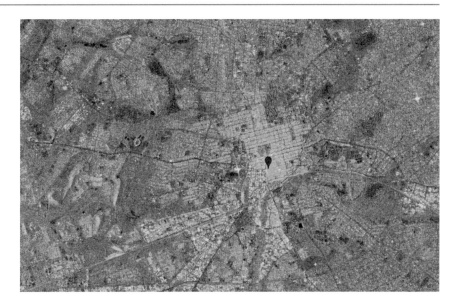

```
//Display the composite image
Map.addLayer(composite, {bands: ['VH', 'VV', 'Ratio'], min: [-30, -25, 0], max: [-5, 0, 15]}, 'SI RGB Composite
RGBMarch', false)
```

Finally, export the Sentinel-1 image polarization so that it can be displayed in GIS software or other software. We can export the composite image GEE Asset folder for use within GEE or to your personal Google Drive or Google Cloud Storage accounts. All exports are sent to the '**Tasks**' tab in the upper right panel.

First create mean and median Sentinel-1 imagery for export.

```
// Create mean SI imagery for export.
var rainy_mean_ASC_VV = collection_S1.select('VV').mean();
print('Mean rainy season VV',rainy_mean_ASC_VV);

var rainy_mean_ASC_VH = collection_S1.select('VH').mean();
print('Mean rainy season VH',rainy_mean_ASC_VH);

// Create median SI imagery for export.
var rainy_median_ASC_VV = collection_S1.select('VV').median();
print('Median rainy season VV',rainy_median_ASC_VV);

var rainy_median_ASC_VH = collection_S1.select('VH').median();
print('Median rainy season VH',rainy_median_ASC_VH);
```

Next, export the mean and median Sentinel-1 imagery for VV and VH.

```
// Export S-1 mean ascending mode VV polarization imagery.
Export.image.toDrive({
 image: rainy_mean_ASC_VV,
 description: 'Hre_S1_mean_Asc_VV_Jan_Mar_20',
 region: boundary,
 maxPixels: 1066537920,
 crs: "EPSG:32736", // EPSG:32736 - WGS 84 / UTM zone 36S - Projected
 scale: 10
});
```

```
// Export S-1 mean ascending mode VH polarization imagery.
Export.image.toDrive({
  image: rainy_mean_ASC_VH ,
  description: 'Hre_S1_mean_Asc_VH_Jan_Mar_20',
  region: boundary,
  maxPixels: 1066537920,
  crs: "EPSG:32736", // EPSG:32736 – WGS 84 / UTM zone 36S – Projected
  scale: 10
});

// Export S-1 median ascending mode VV polarization imagery.
Export.image.toDrive({
  image: rainy_median_ASC_VV,
  description: 'Hre_S1_median_Asc_VV_Jan_Mar_20',
  region: boundary,
  maxPixels: 1066537920,
  crs: "EPSG:32736", // EPSG:32736 – WGS 84 / UTM zone 36S – Projected
  scale: 10
});

// Export S-1 median ascending mode VH polarization imagery.
Export.image.toDrive({
  image: rainy_median_ASC_VH,
  description: 'Hre_S1_median_Asc_VH_Jan_Mar_20',
  region: boundary,
  maxPixels: 1066537920,
  crs: "EPSG:32736", // EPSG:32736 – WGS 84 / UTM zone 36S – Projected
  scale: 10
});
```

1.4 Summary

This chapter introduced geospatial machine learning, the study area, and Sentinel-1 and Sentinel-2 data. Furthermore, hands-on lab exercises on preliminary image processing in GEE were done. Detailed steps were provided on how to access and select Sentinel-1 and Sentinel-2 imagery collection in GEE based on specified criteria. Furthermore, brief discussions were provided on Sentinel-1 orbit and polarization. Sentinel-1 and Sentinel-2 composites were generated and exported for use in GIS software.

1.5 Additional Exercises

i. Export Sentinel-2 imagery for the post-rainy (April 1–June 30, 2020) and dry season (July 1–October 30, 2020) periods.

ii. Compare the Sentinel-2 imagery for the rainy, post-rainy, and dry seasons.

iii. Explain briefly the seasonal changes that are appear in the Sentinel-2 imagery for the rainy, post-rainy, and dry seasons

iv. Export mean and median Sentinel-1 imagery (VV and VH polarization) for the post-rainy (April 1–June 30, 2020) and dry season (July 1–October 30, 2020) periods.

References

Bryan ML (1979) The effect of radar azimuth angle on cultural data. Photogram Eng Remote Sens 45(8):1097–1107

CEOS (2018) A layman's interpretation guide to L-band and C-band synthetic aperture radar data. Available online: http://ceos.org/document_management/SEO/DataCube/Laymans_SAR_Interpretation_Guide_2.0.pdf. Accessed 1 Oct 2020

Christopher AJ (1977) Early settlement and the cadastral framework. In: Kay G, Smout M (eds) Salisbury: a geographical survey of the Capital of Rhodesia, Hodder & Stoughton

Conitz MW (2000) GIS applications in Africa: introduction. Photogram Eng Remote Sens 66:672–673

Eckardt R, Urbazaev M, Salepci N, Pathe C, Schmullius C, Woodhouse I, Stewart C (2019) ECHOES IN SPACE: introduction to radar remote sensing. European Space Agency and EO College

ESA (2015) Sentinel-2 user handbook. https://doi.org/10.13128/REA-22658. Accessed 1 Aug 2020

ESA (2020) SAR formats. Available at https://earth.esa.int/web/sentinel/technical-guides/sentinel-1-sar/products-algorithms/level-1-product-formatting. Accessed 19 Oct 2020

Forkuor G, Cofie O (2011) Dynamics of land-use and land-cover change in Freetown, Sierra Leone and its effects on urban and peri-urban agriculture—a remote sensing approach. Int J Remote Sen 32(4):1017–1037

Gamanya R, De Maeyer P, De Dapper M (2009) Object-oriented change detection for the City of Harare, Zimbabwe. Expert Syst Appl 36(1):571–588

Goldblatt R, Deininger K, Hanson G (2018) Utilizing publicly available satellite data for urban research: mapping built-up land cover and land use in Ho Chi Minh City, Vietnam. Dev Eng 3:83–99

Gorelick N, Hancher M, Dixon M, Ilyushchenko S, Thau D, Moore R (2017) Google earth engine: planetary-scale geospatial analysis for everyone. Remote Sens Environ 202:18–27

Griffiths P, Hostert P, Gruebner O, Linden S (2010) Mapping megacity growth with multi-sensor data. Remote Sens Environ 114(2):426–439

Hardaway G, Gustafs GC, Lichy D (1982) Cardinal effect on Seasat images of urban areas. Photogram Eng Remote Sens 48(3):399–404

Infrastructure and Cities for Economic Development (2017) Zimbabwe's changing urban landscape: evidence and insights on Zimbabwe's urban trends. Available at: https://assets.publishing.service.gov.uk/media/59521681e5274a0a5900004a/ICED_Evidence_Brief_-_Zimbabwe_Urban_Trends_-_Final.pdf. Accessed 19 Oct 2020

Jensen JR (2000) Remote sensing of the environment: an Earth resource perspective. Pearson Education

Kamusoko C, Gamba J, Murakami H (2013) Monitoring urban spatial growth in Harare Metropolitan province, Zimbabwe. Adv Remote Sens 2:322–331

Kamusoko C, Kamusoko OW, Enos C, Gamba J (2021) Mapping urban and peri-urban land cover in Zimbabwe: challenges and opportunities. Geomatics 1:114–147

Ma L, Liu Y, Zhang X, Ye Y, Yin G, Johnson BA (2019) Deep learning in remote sensing applications: a meta-analysis and review. ISPRS J Photogram Remote Sens 152:166–177

Mbiba B (2017) On the periphery: missing urbanisation in Zimbabwe. Available at: https://www.africaresearchinstitute.org/newsite/publications/periphery-missing-urbanisation-zimbabwe/. Accessed 1 Sept 2020

Mundia CN, Aniya M (2005) Analysis of land use/cover changes and urban expansion of Nairobi City using remote sensing and GIS. Int J Remote Sen 26(13):2831–2849

Nemmour H, Chibani Y (2006) Multiple support vector machines for land cover change detection: an application for mapping urban extensions. ISPRS J Photogram Remote Sens 61:125–133

Olofsson P, Herold M, Stehman SV, Woodcock CE, Wulder MA (2014) Good practices for estimating area and assessing accuracy of land change. Remote Sens Environ 148:42–57

Pal M, Mather PM (2003) An assessment of the effectiveness of decision tree methods for land cover classification. Remote Sens Environ 86(4):554–565

Patel D (1998) Some issues of urbanisation and development in Zimbabwe. J Soc Dev Afr 3(2):17–31

R Development Core Team (2005) R: a language and environment for statistical computing. R Foundation for Statistical Computing. Available online at: http://r-project.kr/sites/default/files/2%EA%B0%95%EA%B0%95%EC%A2%8C%EC%86%8C%EA%B0%9C_%EC%8B%A0%EC%A2%85%ED%99%94.pdf. Accessed 3 April 2014

Rodriguez-Galiano VF, Chica-Olmo M, Abarca-Hernandez F, Atkinson PM, Jeganathan C (2012) Random Forest classification of Mediterranean land cover using multi-seasonal imagery and multi-seasonal texture. Remote Sens Environ 121:93–107

Sabins FF (1997) Remote sensing: principles and interpretation. W. H. Freeman and Company, pp 177–240

Seto KC, Fragkias M, Güneralp B, Reilly MK (2011) A meta-analysis of global urban land expansion. PLoS ONE 6(8):e23777. https://doi.org/10.1371/journal.pone.0023777

Stefanov WL, Ramsey MS, Christensen PR (2001) Monitoring urban land cover change: an expert system approach to land cover classification of semiarid to arid centers. Remote Sens Environ 77(2):173–185

UNHABITAT (2008) The state of African cities 2008—a framework for addressing urban challenges in Africa. Nairobi, Kenya

UNHabitat (2020) The strategic plan 2020–2023. Available at: https://unhabitat.org/sites/default/files/documents/2019-09/strategic_plan_2020-2023.pdf. Accessed 1 Sept 2020

United Nations (2018) 68% of the world population projected to live in urban areas by 2050, says UN. Available at: https://www.un.org/development/desa/en/news/population/2018-revision-of-world-urbanization-prospects.html. Accessed 31 Aug

United Nations (2019) World urbanization prospects 2018: highlights. Department of Economic and Social Affairs, Population Division. United Nations Publication. Sales No. E19.XIII.6

United Nations (2020) Goal 11: make cities inclusive, safe, resilient and sustainable. Available online at: https://www.un.org/sustainabledevelopment/cities/. Accessed 1 July 2020

World Health Organization (2020) Urban population growth. Available at: https://www.who.int/gho/urban_health/situation_trends/urban_population_growth_text/en/. Accessed 31 Aug 2020

Xian G, Crane M (2005) Assessments of urban growth in the Tampa Bay watershed using remote sensing data. Remote Sens Environ 97:203–215

Yu X, Wu X, Luo C, Ren P (2017) Deep learning in remote sensing scene classification: a data augmentation enhanced conventional neural network framework. GIS Remote Sens 54(5):741–758

Zerbo A, Delgado RC, González PA (2020) Vulnerability and everyday health risks of urban informal settlements in Sub-Saharan Africa. GHJ 4 (2):46–50

ZimStats (Zimbabwe National Statistics Agency) (2012) Census 2012: preliminary report. Harare, Zimbabwe

Exploratory Analysis and Transformation for Remotely Sensed Imagery

2

Abstract

Exploratory image analysis provides useful insights in order to optimize machine learning models. Furthermore, image transformation—based on spectral and texture indices—can improve land cover classification accuracy. This chapter focuses on creating spectral profiles as well as computing selected spectral and texture indices from Sentinel-1 and Sentinel-2 imagery.

Keywords

Exploratory image analysis • Image transformation • Spectral and texture indices • Sentinel-1 • Sentinel-2 • Land cover classification accuracy

2.1 Introduction

Exploratory image analysis and transformation is critical before running machine learning algorithms or performing image classification. However, this critical step is often underestimated or completely ignored (Kamusoko 2019). Exploratory image analysis encompasses the preparation of training data and the analysis of spectral signatures from the remotely sensed imagery. Image transformation refers to processing of spectral bands into "new" image features. Exploratory image analysis techniques are useful for providing first insights into the underlying structure of the remotely sensed imagery. In addition, exploratory image analysis can be used to select appropriate machine learning parameters and optimize machine learning models. The transformation of remotely sensed imagery into "new" images based on spectral and textural indices improves land cover classification (Tso and Mather 2001; Berberoglu et al. 2000). Shaban and Dikshit (2001) reported that texture indices improve land cover classification accuracy by 9–17%.

In this chapter, we are going to prepare training data using Google Satellite in QGIS. Following that, we are going to create spectral plots based on spectral signatures derived from the training data and Sentinel-2 imagery. In addition, Sentinel-2 imagery will be used to compute selected spectral and texture indices, while Sentinel-1 SAR imagery will be used to compute selected texture indices. Following is a brief discussion on preparing training data sets and performing image transformation.

Electronic supplementary material

The online version of this chapter (https://doi.org/10.1007/978-981-16-5149-6_2) contains supplementary material, which is available to authorized users.

2.1.1 Preparing Training Areas

Nonparametric supervised classification classifiers such as random forest require accurate and reliable training data sets, which contain spectral reflectance or backscatter statistical information derived from satellite imagery (Mather and Koch 2011). In this workbook, training data refers to sample data collected from the field or high resolution imagery. Training data is used to train and validate (test) machine learning classifiers, while independent reference data is used for land cover accuracy assessment. It should be emphasized that the effectiveness of supervised machine learning classifiers and the accuracy of land cover maps is mainly influenced by the availability of accurate and reliable training data. Therefore, training data must be large and representative. The size of the training data is related to the number of variables (e.g., spectral bands or texture variables), the number of target land cover classes, the degree of variability present in the land cover classes, and the requirements of the machine learning classifier (e.g., the number of parameters or weights).

The preparation of training data sets for land cover classification is very challenging given the wide range of satellite data (e.g., hyperspectral versus multispectral imagery) that is available, the study or project purpose, and other factors. In this workbook, we are not going to cover all aspects related to preparing training data sets. Rather, our focus is on preparing training data sets for low-dimensional satellite data such as Sentinel-1, Sentinel-2 and Landsat imagery as well as for nonparametric supervised machine learning classifiers. It is noteworthy to remember that hyperspectral data requires more samples to avoid the curse of high dimensionality. It should also be noted that nonparametric supervised machine learning classifiers operate directly on the training data. Therefore, these classifiers are strongly influenced by the size and accuracy of training samples. Following are some of the suggestions for preparing reliable and accurate training data sets based on literature review.

i. Number of training data samples

In general, it is difficult to determine the appropriate number of training data samples required for land cover classification (Mather and Koch 2011). Researchers consider land cover classes, input variables (spectral bands), machine learning model parameters, and project requirements as a basis to determine the number of training data samples. For example, Mather and Koch (2011) recommend to use a minimum size of $30\times$ number of wavebands for low-dimensional satellite data such as Landsat or Sentinel-2. For example, a Landsat 5 Thematic Mapper (TM) imagery with seven bands would require 210 training samples per land cover class.

ii. Separation of training data used for model training and testing

Training data used to train machine learning models should be separate from test data, which is used for model validation. For remote sensing, training data samples should not be affected by spatial autocorrelation effects. The degree of autocorrelation will depend on (i) the natural association between adjacent pixels, (ii) the pixel dimensions, and (iii) the effects of any data preprocessing. According to Mather and Koch (2011), greater attention should be given to collection of training and test data that represent the range of land surface variability at the satellite image resolution. This is because the distribution of the training data in the feature space determines the positions of the decision boundaries (Tso and Mather 2001). The spatial scale of the landscape features relative to the satellite image resolution affects machine learning model and land cover classification accuracy.

iii. Class imbalance

One of the main challenges of land cover classification is that machine learning classifiers are sensitive to class balance within the training data (Elmes et al. 2020). This is because the number of land cover class instances varies in training data. According to Mellor et al. (2015), the accuracy of random forest-generated land cover maps decreased between 12 and 23% when the training data class balances were highly skewed. In general, a high class imbalance characterized by one or two dominant land cover in the training data represent will result in classification errors. Therefore, it is important to ensure that each target land cover class is well represented in the training data. This is very difficult in practice. However, efforts should be made to collect more samples of the under-represented land cover classes. Alternatively, it is recommended to use data augmentation to create new samples for under-represented land cover classes.

2.1.2 Land Cover Classification Scheme

It is important to clarify some key terms before proceeding. Land cover is the observed biophysical cover on the Earth's surface, while land use refers to human–environment interaction that is characterized by human activity. For example,

Table 2.1 Land cover classification scheme

Land cover	Description
Built-up	Residential, commercial, services, industrial, transportation, communication and utilities, and construction sites
Bareland	Bare sparsely vegetated area with >60% soil background. Includes sand and gravel mining pits, rock outcrops
Cropland	Cultivated land or cropland under preparation, fallow crop land, and cropland under irrigation
Grass/open areas	Grass cover, open grass areas, golf courses, and parks
Woodlands	Woodlands, riverine vegetation, shrub, and bush
Water	Rivers, reservoirs, and lakes

grass/open areas is a land cover class, while a golf course is a land use type. In this workbook, the focus is on mapping land cover. Table 2.1 shows the target land cover classes, which are based on the "Forestry Commission (Zimbabwe) and DSG national woody cover classes" classification schemes, and the author's a priori knowledge of the test sites.

2.1.3 Spectral and Texture Indices

2.1.3 (i). Spectral indices

Spectral indices are derived surface reflectance from two or more wavelengths, which indicate relative abundance of features of interest. While vegetation indices such as the normalized difference vegetation index (NDVI) and soil-adjusted vegetation index (SAVI) are the most common, there are other spectral indices that indicate bare areas, man-made (built-up) features and water. In this chapter, NDVI, the optimized soil-adjusted vegetation index (OSAVI), modified normalized difference water index (MNDWI), and normalized difference built-up index (NDBI) will be computed from Sentinel-2 imagery.

The normalized difference vegetation index (NDVI) is a ratio between the red (R) and near-infrared (NIR) values (Rouse et al. 1974). NDVI is used to quantify vegetation greenness and health. Rondeaux et al. (1996) developed OSAVI, which is based on the SAVI (that is, the ratio between the red and near-infrared (NIR) values with a soil brightness correction factor). OSAVI is based on a standard value of 0.16 for the canopy background adjustment factor in order to account for greater soil variation in low vegetation cover, while increasing sensitivity to vegetation cover greater than 50% (Rondeaux et al. 1996). The index is selected because it best suited for areas with relatively sparse vegetation where soil is visible through the canopy.

Xu (2006) developed MNDWI to enhance open water features while suppressing noise from built-up land, vegetation, and soil. The MNDWI is calculated using the green and shortwave infrared band 1 (SWIR1) bands. Open water has positive values since it absorbs more SWIR wavelengths, while built-up and soil and vegetation features have negative values as soil reflects more SWIR wavelengths. Note that MNDWI is a modification of the NDWI (McFeeters 1996), which was originally developed for use with Landsat TM bands 2 and 5. However, MNDWI can be calculated from other multispectral imagery with green and SWIR bands.

Zha et al. (2003) developed the NDBI, which is calculated from the NIR and shortwave SWIR1 bands. In principal, NDBI indicates a higher reflectance in the SWIR compared to the NIR (Zha et al. 2003) in urban areas. This might work in dense built-up areas or highly developed urban areas. Therefore, NDBI should be interpreted with caution in sparsely built-up areas or newly developed urban areas, especially in developing countries.

2.1.3 (ii). Texture Analysis

Image texture refers to the spatial distribution of tonal variations or the pattern of spatial relationships among the gray levels of neighboring pixels (Haralick et al. 1973). Gray-level co-occurrence matrix (GLCM) tabulates how often different combinations of pixel brightness values (gray levels) occur in an image (Herold et al. 2003). In this chapter, we are going to compute second-order GLCM indices such as the mean, variance, homogeneity, and entropy from Sentinel-1 imagery based on a 3 × 3 moving window. The second-order measures consider the relationship between groups of two neighboring pixels in the original image.

The mean shows the overall intensity level in the neighborhood, while variance measures heterogeneity (Herold et al. 2003). For example, variance increases when the gray-level values differ from their mean. Homogeneity measures image uniformity or similarity, whereas entropy measures the randomness of the intensity distribution. Entropy is high when an image texture is not uniform (Herold et al. 2003; Haralick et al. 1973). Although there are many GLCM indices, most of the indices are highly correlated. Therefore, we are going to compute only four GLCM indices. However, readers are encouraged to test other GLCM indices in their study areas.

2.2 Exploratory Data Analysis and Image Transformation

The following labs will focus on preparing training data, visualizing spectral plots for the rainy season Sentinel-2 Imagery, and computing spectral and texture indices.

2.2.1 Lab 1. Preparing Training Data

Objective

- Digitize polygon training areas using Google Earth imagery and Sentinel-2 imagery in QGIS

Procedure

In this lab, we are going to prepare training data samples using Google Earth and Sentinel-2 imagery in QGIS and create spectral plots in R. The target land cover classes are built-up, bareland, cropland, grass/open areas, woodlands, and water (Tables 2.1 and 2.2). We are going to digitize polygon training samples in QGIS. This exercise requires Internet connection since we are going to digitize polygon training samples from Google Satellite (or ESRI Satellite) available in QGIS.

Table 2.2 Land cover classes, field photos, Sentinel-2, and Google Satellite imagery

Land cover	Field photo	Sentinel-2	Google Satellite
Built-up			
Bare areas			
Cropland			
Grass/open areas			
Woodlands			

Fig. 2.1 Sentinel-2 imagery displayed as bands 8, 4, 3 (RGB) in QGIS

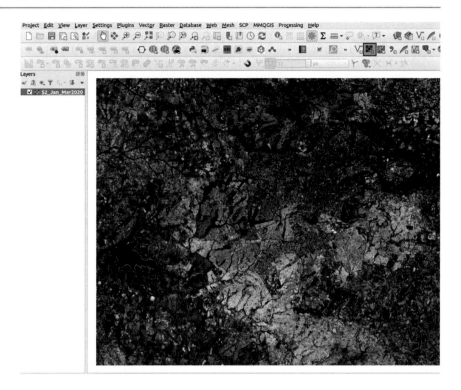

1. Click *Layer* → *Add Layer* → *Add Raster Layer* or simply click the *Add Raster Layer* icon ▪ (Fig. 2.1) to load the Sentinel-2 imagery (S2_Jan_Mar_2020.tif). The imagery is displayed as band 8 (NIR), band 4 (Red), and band 3 (Green).
2. Next, navigate to **Browser** and double-click on **Google Satellite** under the **XYZ Tiles** panel (Fig. 2.2). **Google Satellite** will appear.

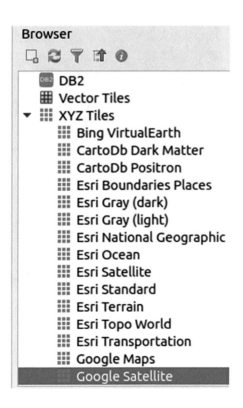

Fig. 2.2 Google Satellite in QGIS

3. Compare Sentinel-2 imagery and **Google Satellite** imagery. You can turn the Sentinel-2 imagery in the *Layers* panel on and off (Fig. 2.3).

4. Create a new empty layer that will store the digitized features. Click on *Layer → Create Layer → New Shapefile Layer* (Fig. 2.4).

5. A *New Shapefile Layer* window appears. Type "training_data.shp" (without quotation marks) under *File Name*. Then, select "*Polygon*" under *Geometry type*. Choose a file path, and click *Save*. You will see a new shapefile in the *Layers Panel* of QGIS.

6. Next, choose "WGS84/UTM zone 36S" (EPSG: 32736) as the coordinate reference system. Finally, enter "Class" (without quotation marks) under *Name*. Keep *Type* as "Text data" and 20 under *Width*. Click *Add to Fields List* button and then *OK*. You are now ready to digitize (Fig. 2.5).

7. Navigate to *View* and then *Toolbars*, and make sure *Digitizing Toolbar* is checked on (Fig. 2.6).

8. Next, right-click on "Training_data" layer, and select *Toggle Editing* (Fig. 2.7).

9. The buttons in the toolbar will be enabled now. To digitize a feature, click on the *Add Polygon Feature* button.

10. Digitize a feature by clicking on the edge of the visible built-up outline (that is built-up area or building). Keep clicking till the polygon is complete. Right click to join the last node to the first one and close the polygon.

11. A dialog will pop-up asking for attribute information. Enter *Id* as 1 and *Class* as "Built-up" of the feature you have just digitized, and click *OK* (Fig. 2.8).

12. Once you are done, click on *Toggle Editing* button. In the pop-up dialog, click *Yes* to save your edits. The digitized feature appears (Fig. 2.9).

13. Repeat steps 8–11 to digitize other land cover training areas (bare areas, cropland, grass/ open areas, woodlands, and water).

14. You can also color the polygons during editing based on the "classes" attribute, which makes it easier for you to estimate the class distribution.

Fig. 2.3 Comparison between Sentinel-2 (left) and Google Satellite imagery (right). Note that the images are displayed at scale of 1:20,000

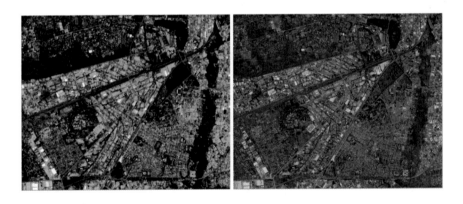

Fig. 2.4 Create a layer and new shapefile

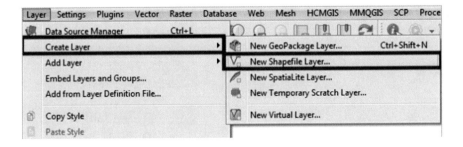

Fig. 2.5 New shapefile layer properties

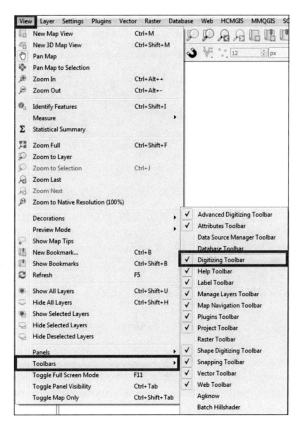

Fig. 2.6 Turn on the digitizing toolbar

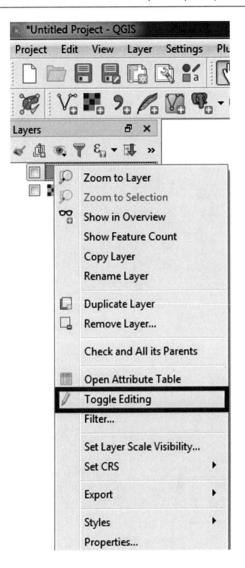

Fig. 2.7 Turn on *Toggle Editing*

15. Double-click the shapefile in the *Layers Panel,* and navigate to the *Style* tab. Ensure that your attribute "classes" is selected in the drop-down menu below.
16. Click *Classify* once to apply an individual color to each class (click on the colored boxes in order to change the colors), and confirm by pressing *OK.*
17. If you think you have collected enough samples, save everything by clicking on the *Toggle Editing* icon again and choose to *Save.*
18. You can also edit the shape of a polygon with the *Node* tool. Delete any unwanted polygons by clicking on the tool called "*Select Features by Area or Single Click*". Once activated, you can left-click on polygons you want to delete, causing them to turn yellow. Then, press the delete key on your keyboard to remove the polygons (only in editing mode).

Tips for preparing training data in QGIS: *Make sure that the polygons for each land cover class are well-distributed within the scene. It is also better to digitize more small polygons than a small number of large homogeneous features. This is because large polygons do not add much value in terms of characterization of the spectral properties of a given feature. It should be noted that preparing training samples is a very important step for land cover classification. Therefore, you should take your time to digitize polygon training samples.*

Fig. 2.8 Built-up feature attributes

Fig. 2.9 Digitized built-up area
displayed on Sentinel-2 and
Google Satellite imagery

2.2.2   Lab 2. Creating Spectral Plots

Objectives

- Create spectral plot in R

Procedures

Understanding spectral reflectance from different land cover classes is important. We are going to prepare a script in R in order to create spectral plots. Note that we can reuse or share script so that others can create spectral plots. Following are the steps for creating a spectral plot of the target land cover classes from the Sentinel-2 imagery. Note, this may take a bit of time depending on your computer memory. Be patient!!!

Step 1: Load the necessary packages or libraries

First, load all required packages. Note, use the *install()* command if some of the packages are not installed on your computer.

```
> # Load the following libraries
> require(sp)
> require(rgdal)
> require(raster)
> require(ggplot2)
```

Step 2: Set up the working directory

Next, set up the working directory where all data sets are located. The data sets consist of Sentinel-2 imagery (that is, raster data in "tif" format) and reference training data polygons (in shapefile format).

```
> # Set up the working directory:
> setwd("/home/kamusokoSentinel-2/Sentinel-2_All_Bands")
```

Step 3: Load the Sentinel-2 imagery

Create a "rvars" (that is, raster variable) object that will contain the rainy season Sentinel-2 imagery. Remember in Chap. 1, you compiled rainy season Sentinel-2 imagery in GEE and exported it as "S2_Jan_Mar2020.tif".

```
> # Import Sentinel-2 imagery.
> rvars <- brick("S2_Jan_Mar2020.tif")
```

Check the attributes of the "rvars" object.

```
> rvars
class         : RasterBrick
dimensions    : 4806, 5950, 28595700, 9  (nrow, ncol, ncell, nlayers)
resolution    : 10, 10  (x, y)
extent        : 260350, 319850, 8001640, 8049700  (xmin, xmax, ymin, ymax)
crs           : +proj=utm +zone=36 +south +datum=WGS84 +units=m +no_defs +ellps=WGS84 +towgs84=0,0,0
source        : Projects/DL_ML_Apps/GEE_Classifications/Sentinel-2/Sentinel-2_All_Bands/S2_Jan_Mar2020.tif
names         : S2_Jan_Mar2020.1, S2_Jan_Mar2020.2, S2_Jan_Mar2020.3, S2_Jan_Mar2020.4, S2_Jan_Mar2020.5
```

Step 4: Load the training data

Next, import the reference training data ("Hre_Revised_Polygon_TA_2020.shp").

```
> # Import training_data shapefile.
> ta_data <- shapefile("Hre_Revised_Polygon_TA_2020.shp")
```

Check the attributes of the "ta_data", object.

```
# Check the attributes of the vector object.
> ta_data
class         : SpatialPolygonsDataFrame
features      : 3352
extent        : 260276.4, 319789.7, 8001727, 8049687  (xmin, xmax, ymin, ymax)
crs           : +proj=utm +zone=36 +south +datum=WGS84 +units=m +no_defs +ellps=WGS84 +towgs84=0,0,0
variables     : 1
names         :    Class
min values    : Bare areas
max values    : Woodlands
```

The "ta_data" object contains 3352 polygon features with the attribute "Class", which consist of target classes such as "Bare areas", "Built-up", and "woodlands".

Next, plot Sentinel-2 and reference training data (Fig. 2.10). Use the argument ***add = TRUE*** in second line to overlay several data layers.

```
> # Plot the training data and Sentinel-2 imagery
> compareCRS(ta_data, rvars)
> plotRGB(rvars, r=7, g=3 , b=2, stretch="lin")
> plot(ta_data, col="yellow", add=TRUE)
```

Convert the class column in our training data set ("ta_data") into factor data type because classifiers work with integer values instead of text such as "Built-up" or "Woodlands".

```
> levels(as.factor(ta_data$Class))
[1] "Bare areas" "Built-up"  "Cropland" "Grass/ open" "Water"     "Woodlands"
```

Next, use the *levels()* function to combine all class instances into a factor-formatted vector.

Fig. 2.10 Training areas (in yellow) displayed on Sentinel-2 imagery

```
> for (i in 1:length(unique(ta_data$Class))) {cat(paste0(i, " ", levels(as.factor(ta_data$Class))[i]),
sep="\n")}
1 Bare areas
2 Built-up
3 Cropland
4 Grass/ open
5 Water
6 Woodlands
```

Rename the spectral bands into the actual Sentinel-2 bands names.

```
> #Check the names of the Sentinel-2 bands
> names(rvars)
[1] "S2_Jan_Mar2020.1" "S2_Jan_Mar2020.2" "S2_Jan_Mar2020.3" "S2_Jan_Mar2020.4" "S2_Jan_Mar2020.5"

> # Rename the Sentinel-2 bands
> names(rvars) <- c('B2', 'B3', 'B4', 'B5', 'B6', 'B7', 'B8', 'B11', 'B12')
```

Remember that in Chap. 1, we exported Sentinel-2 imagery that contains the following nine bands: band 2 (blue), band 3 (green), band 4 (red), band 5 (vegetation red edge), band 6 (vegetation red edge), band 7 (vegetation red edge), band 8 (NIR), band 11 (SWIR), and band 12 (SWIR).

Next, extract raster values. All input raster values from the Sentinel-2 bands and the class values (e.g., bare areas, built-up) of every single pixel within our training polygons should be assembled into a data frame. The *extract()* command from the raster package will extract the raster values. Note that the use of the argument *df = TRUE* guarantees that the output is a data frame. We are also going to measure the time for executing the command.

```
> ## Extract the spectral raster values
> timeStart <- proc.time()
> sp_Sent2 <- extract(rvars, ta_data, df = TRUE)
   user  system elapsed
user  system elapsed
1110.914  12.078 1122.881
```

It took about 19 min to extract the raster values. So be patient!! Remember, this depends on the spatial resolution of the satellite imagery, the area covered by the polygon training data, and computer specifications. Note that the created dataframe (sp_Sent2) has a "ID" column, which contains the Ids of the polygons for each pixel. Therefore, we need to know which polygon (that is, ID) belongs to a specific land cover class. This requires establishing a relationship between each pixel ID and the land cover class using the *match()* function. We use this to add another column to our data query describing each land cover class. Then, we will delete the ID column because we do not need it anymore.

```
> sp_Sent2$cl <- as.factor(ta_data$Class[match(sp_Sent2$ID, seq(nrow(ta_data)))])
> sp_Sent2 <-sp_Sent2[-1]
> summary(sp_Sent2$cl)
Bare areas   Built-up   Cropland Grass/ open     Water  Woodlands
     5780       8752       225597        5088      6846       9238
> str(sp_Sent2)
'data.frame':   261301 obs. of  10 variables:
$ B2 : num  0.1108 0.0951 0.104 0.1128 0.0991 ...
$ B3 : num  0.141 0.13 0.139 0.15 0.132 ...
$ B4 : num  0.172 0.153 0.169 0.181 0.154 ...
$ B5 : num  0.169 0.173 0.197 0.197 0.173 ...
$ B6 : num  0.23 0.197 0.221 0.221 0.197 ...
$ B7 : num  0.253 0.21 0.238 0.238 0.21 ...
$ B8 : num  0.24 0.226 0.24 0.26 0.221 ...
$ B11: num  0.302 0.321 0.351 0.351 0.321 ...
$ B12: num  0.262 0.289 0.318 0.318 0.289 ...
$ cl : Factor w/ 6 levels "Bare areas","Built-up",..: 2 2 2 2 2 2 2 2 2 2 ...
```

Next, we are going to use the *aggregate()* command to combine all the rows of the same class (**. ~ cl**). The mean function (**FUN = mean**) is specified to calculate the arithmetic mean of those groups, while **na.rm** is declared to remove or ignore all NA values.

```
> ## Plots
> sp_Sent2 <- aggregate(. ~ cl, data = sp_Sent2, FUN = mean, na.rm = TRUE)
```

Finally, we are going to define functions to create the spectral plot for visualization purposes. We start by creating an empty plot.

```
> # plot empty plot of a defined size
> plot(0,
+     ylim = c(min(sp_Sent2[2:ncol(sp_Sent2)]), max(sp_Sent2[2:ncol(sp_Sent2)])),
+     xlim = c(1, ncol(sp_Sent2)-1),
+     type = 'n',
+     xlab = "Sentinel-2 bands",
+     ylab = "Reflectance"
+ )
```

Next, we are going to specify the color for each land cover class as follows.

```
> # define colors for each land cover class.
> mycolors <- c("grey","red", "yellow", "light green", "blue", "green")
> # draw one line for each class
> for (i in 1:nrow(sp_Sent2)){
+ lines(as.numeric(sp_Sent2[i, -1]),
+       lwd = 4,
+       col = mycolors[i]
+  )
+ }
```

```
> # add a grid
> grid()

> # add a legend
> legend(as.character(sp_Sent2$cl),
+      x = "topleft",
+      col = mycolors,
+      lwd = 5,
+      bty = "n"
+ )
```

Run the whole script after you finish typing. This will display the spectral plot as shown in (Fig. 2.11).

Observations Figure 2.11 shows nine bands, which we selected from the original Sentinel-bands in Chap. 1. Note that the band names used in Fig. 2.11 are different from the original Sentinel-2 band names. Band 1 is blue, band 2 is green, band 3 is red, band 4 is vegetation red edge 1, band 5 is vegetation red edge 2, band 6 is vegetation red edge 3, band 7 is near infrared (NIR), band 8 is shortwave infrared (SWIR1), and band 9 is SWIR2. Figure 2.11 shows that the blue, green, red, and vegetation red edge 2 clearly separate bare areas, cropland, and built-up areas. However, the other bands (vegetation red edge 1 and 3, NIR, SWIR1, and SWIR2) show spectral similarity within bare areas, cropland, and built-up areas, indicating the difficulty in separating them based on spectral signatures. Great care should be taken into consideration when interpreting spectral information in Fig. 2.11. The spectral plot values represent only the arithmetic mean of the spectral signature for this specific training data in this study area and during the rainy season in 2020. Therefore, the spectral signature is dependent on the imagery, local conditions, and the quantity and quality of training data in given period. However, the spectral information is important because it helps us visually assess the separability of land cover classes before proceeding with image analysis and classification. Furthermore, we can possibly use the spectral information to optimize machine learning classifiers or selecting optimal bands. It is also important to check spectral plots using post-rainy and dry season Sentinel-2 imagery.

Fig. 2.11 Spectral reflectance values derived from rainy season Sentinel-2 imagery (January to March 2020)

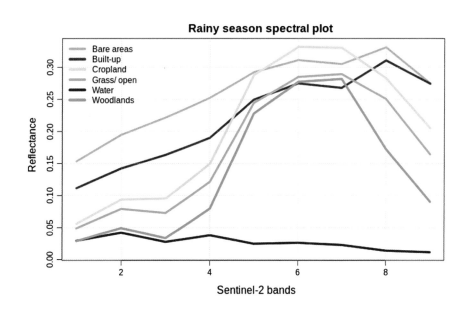

2.2.3   Lab 3. Computing Spectral Indices using Sentinel-2 Imagery

Objective

- Compute spectral indices using Sentinel-2 imagery acquired between January 1, 2020, and March 31, 2020

Procedure

Step 1: Load the necessary packages
First, load the necessary packages.

```
> library(sp)
> library(rgdal)
> library(raster)
> library(ggplot2)
> library(gridExtra)
```

Step 2: Load the Landsat image stack
Set up the working directory

```
setwd("/Projects/Sentinel-2/Sentinel-2_All_Bands/") # in Ubuntu
```

Next, create a list of Sentinel-2 raster layers or bands that will be imported. We are going to import the Sentinel-2 imagery that we compiled in Chap. 1.

```
> rlist=list.files(getwd(),pattern="tif$", full.names=TRUE)
```

Combine or stack the raster layers.

```
> ras_Sent2 <- stack(rlist)
```

Step 3: Check the attributes
Check the attributes and dimensions of the Sentinel-2 imagery.

```
> ras_Sent2
class       : RasterStack
dimensions : 4806, 5950, 28595700, 9 (nrow, ncol, ncell, nlayers)
resolution : 10, 10 (x, y)
extent: 260350, 319850, 8001640, 8049700 (xmin, xmax, ymin, ymax)
crs: +proj=utm +zone=36 +south +datum=WGS84 +units=m +no_defs +ellps=WGS84 +towgs84=0,0,0
names: S2_Jan_Mar2020.1, S2_Jan_Mar2020.2, S2_Jan_Mar2020.3, S2_Jan_Mar2020.4, S2_Jan_Mar2020.5,
S2_Jan_Mar2020.6, S2_Jan_Mar2020.7, S2_Jan_Mar2020.8, S2_Jan_Mar2020.9
```

The "rast_Sent2" (refers to raster Sentinel-2) is a raster object with 4806 rows, 5950 columns, 28,595,700 pixels, and nine bands. The spatial resolution of the raster object is 10 m and has coordinate reference system, which is WGS84 UTM Zone 36 South. Remember that in Chap. 1, we exported Sentinel-2 imagery that contains the following nine bands: band 2 (Blue), band 3 (Green), band 4 (Red), band 5 (Vegetation red edge), band 6 (Vegetation red edge), band 7 (Vegetation red edge), band 8 (NIR), band 11 (SWIR), and band 12 (SWIR). However, the band names changed when we imported the Sentinel-2 imagery. For example, S2_Jan_Mar2020.1 represents band 2 (Blue), while S2_Jan_Mar2020.2 represents band 3 (Green). Note that you can use the *names()* command as we have done in lab 2 to rename the band names if this long name confuses you.

Fig. 2.12 Sentinel-2 single band collection acquired between January 1, 2020, and March 31, 2020

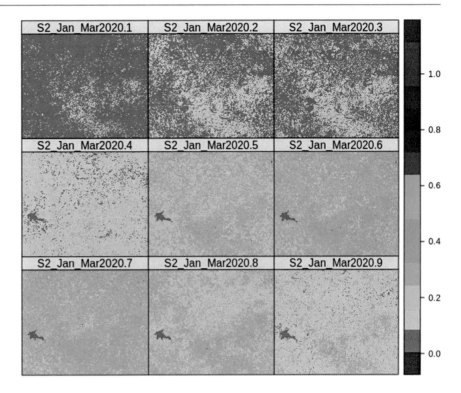

Step 4: Display the Landsat images

Next, display the raster object as single bands (Fig. 2.12) and as a composite imagery (Fig. 2.13).

```
> # Dispaly Sentinel-2 bands.
> spplot(ras_Sent2, col.regions = rainbow(99, start = 1))


> # Display the composite Sentinel-2 subset for Harare.
> plotRGB(ras_Sent2, r=7, g=3, b=2, stretch="lin")
```

Figure 2.12 shows the output, which consists of Sentinel-2 band 2 (S2_Jan_Mar2020.1), band 3 (S2_Jan_Mar2020.2), band 4 (S2_Jan_Mar2020.3), band 5 (S2_Jan_Mar2020.4), band 6 (S2_Jan_Mar2020.5), band 7 (S2_Jan_Mar2020.6), band 8 (S2_Jan_Mar2020.7), band 10 (S2_Jan_Mar2020.8), and band 11 (S2_Jan_Mar2020.9) (2.14).

Step 5: Compute selected vegetation indices

Define a function to compute NDVI, where *img* is a multilayer Raster object, and *i* and *k* are the layer numbers (index) used to compute the vegetation index.

```
> # Create a VI function (vegetation index)
> ndvi <- function(img, k, i) {
+   bk <- img[[k]]
+   bi <- img[[i]]
+   vi <- (bk - bi) / (bk + bi)
+   return(vi)
+ }
```

Next, let use the function to compute NDVI and then check the properties of NDVI object. Note that, Sentinel-2 band 8 (NIR) is indexed as 7, while band 4 (Red) is indexed as 3.

Fig. 2.13 Sentinel-2 false color composite, displayed as Red (band 8), Green (band 4), and Blue (band 3)

Fig. 2.14 Spectral indices for the rainy season (January–March 2020)

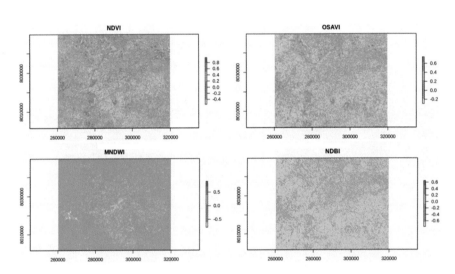

```
# Compute ndvi
> ndvi_rainy_season <- ndvi(ras_Sent2, 7, 3)

> # Check the properties of the NDVI image.
> ndvi_rainy_season
class       : RasterLayer
dimensions  : 4806, 5950, 28595700  (nrow, ncol, ncell)
resolution  : 10, 10  (x, y)
extent      : 260350, 319850, 8001640, 8049700  (xmin, xmax, ymin, ymax)
crs         : +proj=utm +zone=36 +south +datum=WGS84 +units=m +no_defs +ellps=WGS84 +towgs84=0,0,0
source      : memory
names       : layer
values      : -0.6026058, 0.9995222  (min, max)
```

The output shows that NDVI has a range between −0.6 and 1.

We can also create a function to compute OSAVI and then use it for computation.

```
> osavi <- function(img, k, i) {
+   bk  <- img[[k]]
+   bi  <- img[[i]]
+   vi  <- (bk - bi) / (bk + bi + 0.16)
+   return(vi)
+ }


# Compute OSAVI
> osavi_rainy_season <- osavi(ras_Sent2, 7, 3)
```

Check the properties of the OSAVI object.

```
> # Check the properties of the OSAVI image.
> osavi_rainy_season
class       : RasterLayer
dimensions  : 4806, 5950, 28595700  (nrow, ncol, ncell)
resolution  : 10, 10  (x, y)
extent      : 260350, 319850, 8001640, 8049700  (xmin, xmax, ymin, ymax)
crs         : +proj=utm +zone=36 +south +datum=WGS84 +units=m +no_defs +ellps=WGS84 +towgs84=0,0,0
source      : memory
names       : layer
values      : -0.3737614, 0.7507894  (min, max)
```

Note that OSAVI has a range between −0.37 and 0.75, which is different from NDVI.
Next, we going to create a general function, which we will use to compute MNDWI.

```
> # MNDWI = (Green - SWIR) / (Green + SWIR)
> SI <- function(img, k, i) {
+   bk  <- img[[k]]
+   bi  <- img[[i]]
+   vi  <- (bk - bi) / (bk + bi)
+   return(vi)
+ }


# Compute MNDWI
> mndwi_rainy_season <-SI(ras_Sent2, 8, 2)


> # Check the properties of the MNDWI image.
> mndwi_rainy_season
class       : RasterLayer
dimensions  : 4806, 5950, 28595700  (nrow, ncol, ncell)
resolution  : 10, 10  (x, y)
extent      : 260350, 319850, 8001640, 8049700  (xmin, xmax, ymin, ymax)
crs         : +proj=utm +zone=36 +south +datum=WGS84 +units=m +no_defs +ellps=WGS84 +towgs84=0,0,0
source      : memory
names       : layer
values      : -0.8559867, 0.9997192  (min, max)
```

Note that MNDWI has a range between −0.86 and 0.99.

```
> # Save the MNDWI image.
> writeRaster("/home/Projects/mndwi_rainy_season/mndwi_rainy_season_2020.tif", datatype='FLT4S')
```

We can also use the SI function that we created above to compute NDBI.

```
> # compute NDBI, where NDBI = (SWIR - NIR) / (SWIR + NIR)
> # Compute ndbi.
> ndbi_rainy_season <- SI(ras_Sent2, 8, 7)


> # Check the properties of the NDBI image.
> ndbi_rainy_season
class       : RasterLayer
dimensions  : 4806, 5950, 28595700 (nrow, ncol, ncell)
resolution  : 10, 10 (x, y)
extent      : 260350, 319850, 8001640, 8049700 (xmin, xmax, ymin, ymax)
crs         : +proj=utm +zone=36 +south +datum=WGS84 +units=m +no_defs +ellps=WGS84 +towgs84=0,0,0
source      : memory
names       : layer
values      : -0.8623835, 0.8664284 (min, max)
```

Next, display the computed spectral indices (Fig. 2.14). Use the **par**() function to create a matrix of *nrows x ncols* plots.

```
> # Display all spectral indices
> par(mfrow = c(2,2))
> plot(ndvi_rainy_season, col =rev(terrain.colors(6)), main = "NDVI")
> plot(osavi_rainy_season, col =rev(terrain.colors(6)), main = "OSAVI")
> plot(mndwi_rainy_season, col =rev(terrain.colors(6)), main = "MNDWI")
> plot(ndbi_rainy_season, col =rev(terrain.colors(6)), main = "NDBI")
```

Finally, save all spectral indices to file.

```
# Save the NDVI image.
>writeRaster(ndvi_rainy_season, "/Projects/ndvi_rainy_season_2020.tif", datatype='FLT4S', over-
write = TRUE)
>writeRaster(ndvi_rainy_season, "Projects/ndvi_rainy_season_2020.tif", datatype='FLT4S', overwrite = TRUE)
>writeRaster(osavi_rainy_season,"Projects/osavi_rainy_season_2020.tif", datatype='FLT4S', over-
write = TRUE)
>writeRaster(mndwi_rainy_season,"Projects/mndwi_rainy_season_2020.tif", datatype='FLT4S', over-
write = TRUE)
```

2.2.4   Lab 4. Computing Texture Indices using Sentinel-2 Imagery

Objective

- Compute texture indices using rainy season Sentinel-2 imagery

Procedures

We are going to use rainy season Sentinel-2 band 2 (blue), band 3 (green) and band 4 (red), and band 8 (near-infrared) for computing texture indices. This is because these bands originally have a 10 m spatial resolution. Following are the procedures.

Step 1: Install and load the necessary packages or libraries.
First, install the "glcm" package (Zvoleff 2016).

```
> install.packages("glcm")
```

Next, load the following packages.

```
> require(sp)
> require(rgdal)
> require(raster)
> require(glcm)
> require(gridExtra)
```

Step 2: Set up the working directory
Create your working directory and then import the rainy season Sentinel-2 imagery.

```
setwd("Projects/DL_ML_Apps/GEE_Classifications/Urban/Harare/SatelliteImagery/Sentinel/2020/Sentinel-2/
Sentinel-2_All_Bands")
```

Next, import the rainy season Sentinel-2 imagery.

```
> # Import Sentinel-2 imagery for the rainy season (January to March 2020).
> rvars <- brick("S2_Jan_Mar2020.tif")
```

Step 3: Check the attributes

```
> # Check the attributes of the 'rvars' object that we just created.
> rvars
class       : RasterBrick
dimensions  : 4806, 5950, 28595700, 9 (nrow, ncol, ncell, nlayers)
resolution  : 10, 10 (x, y)
extent      : 260350, 319850, 8001640, 8049700 (xmin, xmax, ymin, ymax)
crs         : +proj=utm +zone=36 +south +datum=WGS84 +units=m +no_defs +ellps=WGS84 +towgs84=0,0,0
source      : /Sentinel-2/Sentinel-2_All_Bands/S2_Jan_Mar2020.tif
names       : S2_Jan_Mar2020.1, S2_Jan_Mar2020.2, S2_Jan_Mar2020.3, S2_Jan_Mar2020.4, S2_Jan_Mar2020.5,
S2_Jan_Mar2020.6, S2_Jan_Mar2020.7, S2_Jan_Mar2020.8, S2_Jan_Mar2020.9
```

Step 4: Compute texture index for band 2
Compute the second-order GLCM indices (mean, variance, homogeneity, and entropy) for Sentinel-2 band 2 based on a 3×3 window size.

```
> # Compute the second order GLCM index for Sentinel-2 image band 1.
> glcm_rainy_sent2 <- glcm(raster(rvars, layer=1), window = c(3, 3), statistics = c("mean", "variance", "
homogeneity", "entropy"))
```

Display glcm texture for band 2 (Fig. 2.15).

```
> # Plot glcm texture for band 2
> plot(glcm_rainy_sent2)
```

Save the glcm texture images for band 2. We are going to use the texture images in Chaps. 3 and 5.

```
> # Save the glcm texture for band 2 glcm_sent2_b2).
> writeRaster(glcm_rainy_sent2$glcm_mean,"RS_mean_S2_b2.tif", datatype='FLT4S', overwrite = TRUE)
> writeRaster(glcm_rainy_sent2$glcm_variance,"RS_variance_S2_b2.tif", datatype='FLT4S', overwrite = TRUE)
> writeRaster(glcm_rainy_sent2$glcm_homogeneity,"RS_homogeneity_S2_b2.
```

Fig. 2.15 Rainy season
Sentinel-2 band 2 texture images

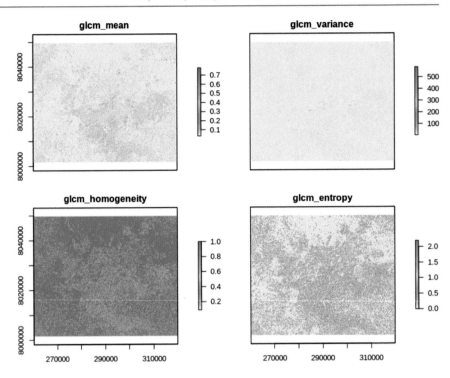

```
tif", datatype='FLT4S', overwrite = TRUE)
> writeRaster(glcm_rainy_sent2$glcm_entropy, "RS_entropy_S2_b2.tif", datatype='FLT4S', overwrite = TRUE)
```

Step 5: Compute texture index for band 3
Compute the second-order GLCM indices for rainy season Sentinel-2 band 3.

```
> # Compute the second order GLCM index for Sentinel-2 image band 3.
> glcm_rainy_sent2 <- glcm(raster(rvars, layer=2), window = c(3, 3), statistics = c("mean", "variance",
"homogeneity", "entropy"))
```

Display glcm texture for band 3 (Fig. 2.16).

```
> # Plot glcm texture for band 3
> plot(glcm_rainy_sent2)
```

Save the glcm texture images for band 3.

```
> # Save the glcm texture for band 3 (glcm_sent2_b3).
> writeRaster(glcm_rainy_sent2$glcm_mean, "RS_mean_S2_b3.tif", datatype='FLT4S', overwrite = TRUE)
> writeRaster(glcm_rainy_sent2$glcm_variance, "RS_variance_S2_b3.tif", datatype='FLT4S', overwrite = TRUE)
> writeRaster(glcm_rainy_sent2$glcm_homogeneity, "RS_homogeneity_S2_b3.tif", datatype='FLT4S', over-
write = TRUE)
> writeRaster(glcm_rainy_sent2$glcm_entropy, "RS_entropy_S2_b3.tif", datatype='FLT4S', overwrite = TRUE)
```

Step 6: Compute texture index for band 4
Compute the second-order GLCM indices for rainy season Sentinel-2 band 4.

```
> # Compute the second order GLCM index for Sentinel-2 image band 4.
> glcm_rainy_sent2 <- glcm(raster(rvars, layer=3), window = c(3, 3), statistics = c("mean", "variance",
"homogeneity", "entropy"))
```

Fig. 2.16 Rainy season
Sentinel-2 band 3 texture images

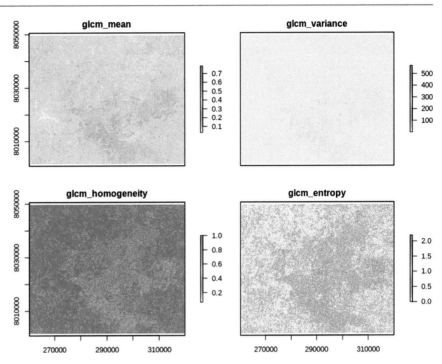

Display glcm texture for band 4 (Fig. 2.17).

```
> # Plot glcm texture for band 4
> plot(glcm_rainy_sent2)
```

Save the glcm texture images for band 4.

```
> # Save the glcm texture for band 4 (glcm_sent2_b4).
> writeRaster(glcm_rainy_sent2$glcm_mean,"RS_mean_S2_b4.tif", datatype='FLT4S', overwrite = TRUE)
> writeRaster(glcm_rainy_sent2$glcm_variance,"RS_variance_S2_b4.tif", datatype='FLT4S', overwrite = TRUE)
> writeRaster(glcm_rainy_sent2$glcm_homogeneity,"RS_homogeneity_S2_b4.tif", datatype='FLT4S',
overwrite = TRUE)
> writeRaster(glcm_rainy_sent2$glcm_entropy, "RS_entropy_S2_b4.tif", datatype='FLT4S', overwrite = TRUE)
```

Step 7: Compute texture index for band 8
Compute the second-order GLCM indices for rainy season Sentinel-2 band 8.

```
> # Compute the second order GLCM index for Sentinel-2 image band 8.
> glcm_rainy_sent2 <- glcm(raster(rvars, layer=7), window = c(3, 3), statistics = c("mean", "variance",
"homogeneity", "entropy"))
```

Display glcm texture for band 8 (Fig. 2.18).

```
> # Plot glcm texture for band 8
> plot(glcm_rainy_sent2)
```

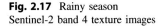

Fig. 2.17 Rainy season
Sentinel-2 band 4 texture images

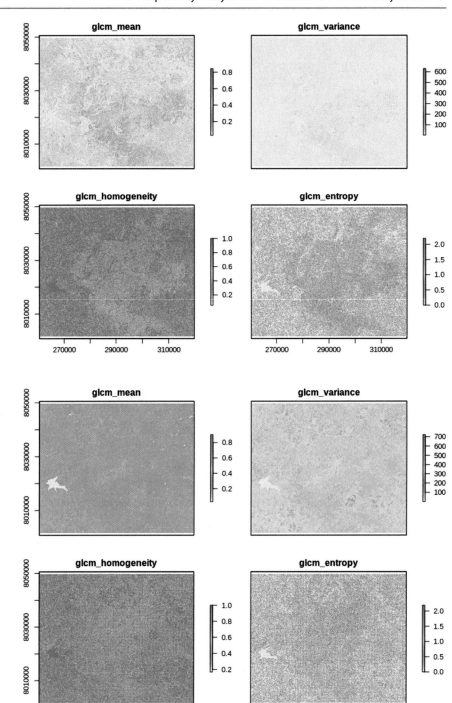

Fig. 2.18 Rainy season
Sentinel-2 band 8 texture images

Save the glcm texture images for band 8.

```
> # Save the glcm texture for band 8 (glcm_sent2_b8).
> writeRaster(glcm_rainy_sent2$glcm_mean,"RS_mean_S2_b8.tif", datatype='FLT4S', overwrite = TRUE)
> writeRaster(glcm_rainy_sent2$glcm_variance,"RS_variance_S2_b8.tif", datatype='FLT4S', overwrite = TRUE)
> writeRaster(glcm_rainy_sent2$glcm_homogeneity,"RS_homogeneity_S2_b8.tif", datatype='FLT4S', over-
write = TRUE)
> writeRaster(glcm_rainy_sent2$glcm_entropy, "RS_entropy_S2_b8.tif", datatype='FLT4S', overwrite = TRUE)
```

2.2.5 Lab 5. Computing Texture Indices using Sentinel-1 Imagery

Objective

- Compute texture indices from rainy season Sentinel-1 imagery VV and VH polarization

Procedures

In this lab, we are going to compute mean, variance, homogeneity, and entropy GLCM texture variables using rainy season Sentinel-1 imagery VV and VH polarization.

Step 1: Install and load the necessary packages or libraries
First, load the following packages.

```
> require(sp)
> require(rgdal)
> require(raster)
> require(glcm)
> require(gridExtra)
```

Step 2: Set up the working directory and import the imagery
Set up the working directory as we have done in the previous labs.

```
setwd("/Projects/DL_ML_Apps/Sentinel_1/Dry_Rainy_Season_2020/")
```

Next, import Sentinel-1 mean VV and VH polarization, and then, display the imagery.

```
> # Import Sentinel-1 mean VV for the rainy season (January to March 2020).
> RS_mean_Asc_VV <- brick("Hre_S1_mean_Asc_VV_Jan_Mar_20.tif")

> # Plot the imagery
> plot(RS_mean_Asc_VV)

> # Import Sentinel-2 mean VH for the rainy season (April to June 2020).
> RS_mean_Asc_VH <- brick("Hre_S1_mean_Asc_VV_Apr_Jun_20.tif")

> # Plot the imagery
> plot(RS_mean_Asc_VH)
```

Create Sentinel-1 false color composite imagery, which comprises mean VV, VH, and VV polarization (Fig. 2.19).

```
> # Combine the imagery
> rvars <- stack(RS_mean_Asc_VV, RS_mean_Asc_VH, RS_mean_Asc_VV)

> # Plot the imagery as rgb
> plotRGB(rvars, r=1, g=2 , b=3, stretch="lin")
```

Step 3: Compute the second-order GLCM index for rainy season Sentinel-1 VV
Compute the second-order GLCM indices (mean, variance, homogeneity, and entropy) for Sentinel-1 VV polarization based on a 3×3 window size.

```
> # Compute the second order GLCM index for rainy season Sentinel-1 mean VV polarization.
> RS_glcm_mean_Asc_VV <- glcm(raster(RS_mean_Asc_VV, layer=1), window = c(3, 3), statistics = c("mean", "variance", "homogeneity", "entropy"))
```

Fig. 2.19 Rainy season
Sentinel-1 composite imagery
(VV, VH, and VV polarization)

Fig. 2.20 Rainy season
Sentinel-1 VV polarization image

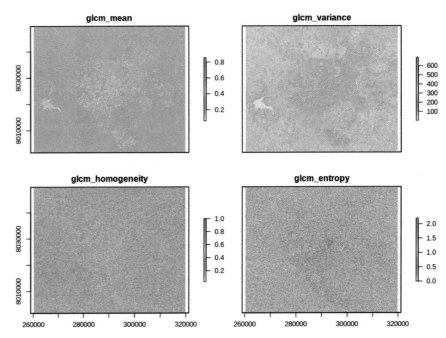

Next, display the raster object as single images (Fig. 2.20).

```
> # Plot glcm texture for mean VV polarization
> plot(RS_glcm_mean_Asc_VV)
```

Save the glcm texture images for VV polarization.

```
> # Save the glcm texture for VV polarization (RS_glcm_mean_Asc_VV).
> writeRaster(RS_glcm_mean_Asc_VV$glcm_mean,"RS_mean_asc_VV.tif", datatype='FLT4S', overwrite = TRUE)

> writeRaster(RS_glcm_mean_Asc_VV$glcm_variance,"RS_variance_asc_VV.tif", datatype='FLT4S',
overwrite = TRUE)
```

```
> writeRaster(RS_glcm_mean_Asc_VV$glcm_homogeneity,"RS_homogeneity_asc_VV.tif", datatype='FLT4S',
overwrite = TRUE)


>writeRaster(RS_glcm_mean_Asc_VV$glcm_entropy,"RS_entropy_asc_VV .tif",
     datatype='FLT4S', overwrite = TRUE)
```

Step 4: Compute the second-order GLCM index for rainy season Sentinel-1 VH

Compute the second-order GLCM indices (mean, variance, homogeneity, and entropy) for Sentinel-1 VH polarization based on a 3×3 window size.

```
> # Compute the second order GLCM index for rainy season Sentinel-1 VH polarization.
> RS_glcm_mean_Asc_VH <- glcm(raster(RS_mean_Asc_VH, layer=1), window = c(3, 3), statistics = c("mean",
"variance", "homogeneity", "entropy"))
```

Next, display the raster object as single images (Fig. 2.21).

```
> # Plot the mean VHpolarization
> plot(RS_glcm_mean_Asc_VH)
```

Save the glcm texture images for VH polarization.

```
> # Save the glcm texture for VH polarization (RS_glcm_mean_Asc_VH).
> writeRaster(RS_glcm_mean_Asc_VH$glcm_mean,"RS_mean_asc_VH.tif", datatype='FLT4S', overwrite = TRUE)
> writeRaster(RS_glcm_mean_Asc_VH$glcm_variance,"RS_variance_asc_VH.
tif", datatype='FLT4S', overwrite = TRUE)


> writeRaster(RS_glcm_mean_Asc_VH$glcm_homogeneity,"RS_homogeneity_asc_VH.tif", datatype='FLT4S',
overwrite = TRUE)


>writeRaster(RS_glcm_mean_Asc_VH$glcm_entropy,"RS_entropy_asc_VH.tif",
     datatype='FLT4S', overwrite = TRUE)
```

Fig. 2.21 Rainy season Sentinel-1 VH polarization image

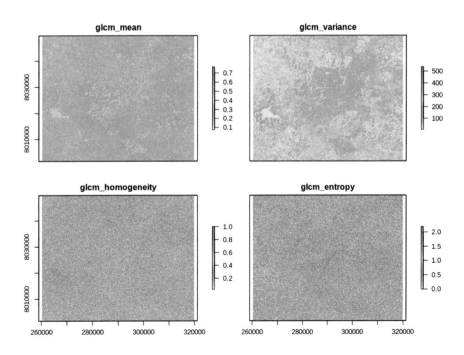

Observations It should be noted that texture varies with the landscape characteristics of the study area and the remotely sensed imagery used. As a result, it is difficult to select the appropriate texture indices for image classification (Chen et al. 2004). This is because remote sensing image analysts have to compute texture indices using different moving window sizes. This is time-consuming. Furthermore, interpretation of texture measures is not intuitive. While we have only computed four texture measures using 3×3 window, it is recommended to compute all texture indices and different window sizes. This may help to improve land cover classification.

2.3 Summary

This chapter covered exploratory analysis and transformation for remotely sensed imagery. We prepared training data and created spectral profile. Spectral indices were computed from Sentinel-2 imagery, while texture indices were computed from both Sentinel-1 and Sentinel-2 imagery.

2.4 Additional Exercises

  i. Create spectral profiles using post-rainy (April 1–June 30, 2020) and dry season (July 1–October 31, 2020) Sentinel-2 imagery.
 ii. Discuss the differences in the spectral profiles computed from the rainy, post-rainy, and dry season Sentinel-2 imagery.
iii. Compute spectral indices for the post-rainy and dry season periods using Sentinel-2 imagery downloaded from GEE.
 iv. Explain briefly the seasonal changes that are appear in the spectral indices for the rainy, post-rainy, and dry seasons.
  v. Compute texture indices for the post-rainy and dry season periods using Sentinel-2 imagery downloaded from GEE.
 vi. Compute texture indices for the post-rainy and dry season periods using Sentinel-1 imagery (VV and VV polarizations).

Tip *Compile and download Sentinel-2 imagery from Google Earth Engine (GEE) using this link:* https://code.earthengine.google.com/83f0a8cddb89ccc55e2268aa9ecb01d9.

References

Berberoglu S, Lloyd CD, Atkinson PM, Curran PJ (2000) The integration of spectral and textural information using neural networks for land cover mapping in the Mediterranean. Comput Geosci 26:385–396

Chen D, Stow DA, Gong P (2004) Examining the effect of spatial resolution and texture window size on classification accuracy: an urban environment case. Int J Remote Sen 25:2177–2192

Elmes A, Alemohammad H, Avery R, Caylor K, Eastman R, Fishgold L, Friedl MA, Pontius RG, Jain M, Kohli D (2020) Accounting for training data error in machine learning applied to Earth observations. Remote Sens 12(1034):1–39

Haralick RM, Shanmugam K, Dinstein I (1973) Textural features for image classification. IEEE Trans Syst Man Cybern SMC-3, pp 610–621

Herold M, Goldstein NC, Clarke KC (2003) The spatiotemporal form of urban growth: measurement, analysis and modelling. Remote Sens Environ 86:286–302

Kamusoko C (2019) Remote sensing image classification in R. Springer

Mather PM, Koch M (2011) Computer processing of remotely-sensed images: an introduction. Wiley-Blackwell, Chichester

McFeeters S (1996) The use of normalized difference water index (NDWI) in the delineation of open water features. Int J Remote Sen 17: 1425–1432

Mellor A, Boukir S, Haywood A, Jones S (2015) Exploring issues of training data imbalance and mislabelling on random forest performance for large area land cover classification using the ensemble margin. ISPRS J Photogram Remote Sens 105:155–168

Rondeaux G, Steven M, Baret F (1996) Optimization of soil-adjusted vegetation indices. Remote Sens Environ 55:95–107

Rouse JW Jr, Haas R, Schell J, Deering DW (1974) Monitoring vegetation systems in the great plains with ERTS. In: Third ERTS symposium, NASA, 351, pp 309–317

Shaban MA, Dikshit O (2001) Improvement of classification in urban areas by the use of textural features: the case study of Lucknow City, Uttar Pradesh. Int J Remote Sens 22(4):565–593

Tso B, Mather PM (2001) Classification methods for remotely-sensed data. Taylor and Francis, New York

Xu H (2006) Modification of normalised difference water index (NDWI) to enhance open water features in remotely sensed imagery. Int J Remote Sens 27(14):3025–3033

Zha Y, Gao J, Ni S (2003) Use of normalized difference built-up index in automatically mapping urban areas from TM imagery. Int J Remote Sens 24(3):583–594

Zvoleff A (2016) Package 'glcm'. Available at https://cran.r-project.org/web/packages/glcm/glcm.pdf. Accessed 22 June 2017

Mapping Urban Land Cover Using Multi-seasonal Sentinel-2 Imagery, Spectral and Texture Indices

3

Abstract

Urban land cover information is a prerequisite for urban land use and environmental and climate change studies. However, urban land cover information is either unavailable or outdated in most developing countries. The availability of free Earth observation (EO) data such as Sentinel-2 coupled with powerful machine learning methods provides opportunities to improve urban land cover mapping. In this chapter, multi-seasonal Sentinel-2 imagery as well as spectral and textural indices will be used to map land cover.

Keywords

Earth observation (EO) data • Multi-seasonal Sentinel-2 imagery • Spectral indices • Texture indices • Machine learning

3.1 Introduction

3.1.1 Background

As I mentioned in Chap. 1, mapping land cover in urban areas is still difficult and challenging (Wania et al. 2014; Schug et al. 2018). This is attributed to a number of factors. First, built-up areas in sparse urban or peri-urban areas appear identical to fallow cropland and bare areas because these features exhibit high reflectance in the visible-infrared wavelengths (Herold et al. 2004; Schneider 2012). For example, spectral similarity between newly developed built-up areas and other land cover surfaces such as fallow cropland fields and bare areas has been reported to cause classification errors, especially in peri-urban areas in developing countries (Kamusoko et al. 2013, 2021; Schug et al. 2018). This is because most built-up structures in peri-urban areas are made of the same materials found in the surrounding areas, which results in low object-to-background contrast (Jensen 2000). Second, urban areas are generally characterized by a variety of surficial materials that vary in size and shape, which result in the mixed pixel problem (Small 2004). The mixed problem is more intense in newly developed peri-urban areas characterized by small and patchy developments (Korn et al. 2009).

The availability of free Sentinel-2 data provides a great opportunity to address some of the key challenges. This is because Sentinel-2 data has high spatial and temporal resolutions. Therefore, spectral–temporal features derived from multi-seasonal Sentinel-2 data can be used to discriminate built-up areas from cropland and bare areas. Past studies have shown that multi-seasonal or multi-temporal Sentinel-2 or Landsat data improve land cover mapping in urban and peri-urban areas (Kamusoko et al. 2021; Yuan et al. 2005). This is because built-up spectral responses are largely persistent during the different seasons, while non-built-up (e.g., cropland and bare areas) spectral responses change (Griffiths et al. 2010). In this chapter, we are going to test the effectiveness of multi-seasonal (rainy, post-rainy, and dry season) Sentinel-2 imagery in separating built-up from non-built surfaces.

3.1.2 Land Cover Mapping Using Multi-seasonal Imagery and Other Derived Data

In this chapter, we are going to use multi-seasonal median Sentinel-2 imagery acquired during the rainy (January–March), post-rainy (April–June), and dry (July–October) seasons (Fig. 3.1). Furthermore, we will test if spectral and texture indices can improve land cover classification. In the rainy season imagery, croplands, grass/open areas, and woodlands are spectrally similar, while asphalt and concrete can be distinguished from bare surfaces in developed urban areas (Fig. 3.1). However, it is difficult to separate cropland areas where land is being prepared for cultivation (bare cropland areas) from built-up areas in the rainy season imagery, especially in developing peri-urban areas. Figure 3.2a shows that the blue, green, red, and vegetation red edge 2 clearly separates bare areas, cropland, and built-up areas. However, the other bands (vegetation red edge 1 and 3, NIR, SWIR1, and SWIR2) show spectral similarity for bare areas, cropland, and built-up areas. This shows the difficulty of separating these land cover classes based only on spectral reflectance (Fig. 3.2a).

The post-rainy season imagery clearly distinguishes woodlands that are leaf-on (Fig. 3.1). However, it is difficult to separate harvest or post-harvest cropland areas (which were originally planted with annual crops such as maize) from grass/open areas. In addition, it difficult to separate built-up surfaces from cropland or bare areas. In the post-rainy season Sentinel-2 imagery, cropland areas that are being prepared for irrigation respond as bare soil, making it difficult to separate them from built-up areas (Fig. 3.1). The post-rainy season spectral profile indicates narrow separability in the blue, green and red bands for built-up and cropland areas, and close spectral similarity in the other bands (Fig. 3.2b). Furthermore, grass/open areas appear spectrally close to cropland areas during the post-rainy season (Fig. 3.2b). This because most crops are harvested during the post-rainy season. As a result, harvest or post-harvest cropland areas covered with crop residue or grass have similar spectral reflectance as grass/open areas.

The dry season imagery shows increased difficulty to separate built-up areas from cropland and bare areas. This is because most cropland are bare during the dry season. Figure 3.2c shows that the spectral reflectance between the built-up and cropland areas is quite close, and therefore, it is difficult to separate these classes. Note that grass/open areas are also spectrally close to cropland areas during the dry season (Fig. 3.2c).

Fig. 3.1 Multi-seasonal Sentinel-2 imagery showing changes in crop fields, bare areas, and built-up areas

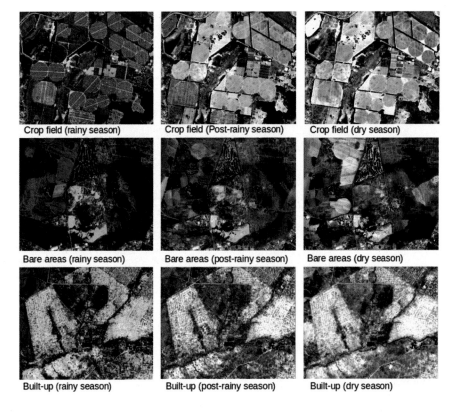

Fig. 3.2 **a** Rainy season spectral
plot computed from rainy season
Sentinel-2 imagery; **b** Post-rainy
season spectral plot computed
from post-rainy season Sentinel-2
imagery and **c** Dry season spectral
plot computed from dry season
Sentinel-2 imagery

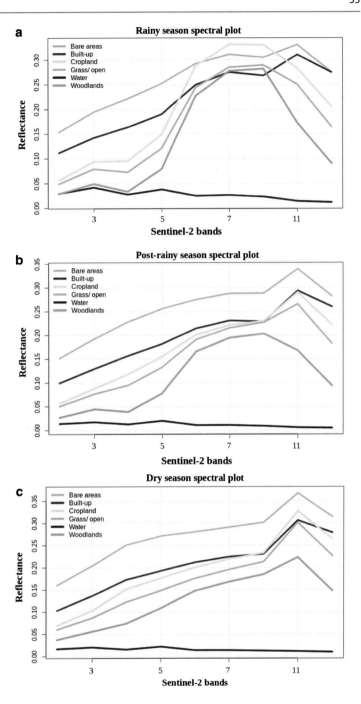

3.2   Land Cover Mapping Labs

This chapter consist of two lab exercises. The first lab exercise focuses on mapping land cover using multi-seasonal
Sentinel-2 imagery (rainy, post-rainy, and dry season). The second lab exercise focuses on mapping land cover using
multi-seasonal Sentinel-2 imagery and spectral and texture indices. The random forest (RF) classifier will be used for all
image classifications. We are going to use spectral and texture indices computed in Chap. 2. The following are the lab
exercises.

3.2.1 Lab 1. Mapping Land Cover Using Multi-seasonal Sentinel-2 Imagery

Objective

- Map land cover using multi-seasonal Sentinel-2 imagery and a random forest classifier

Procedure

In this lab, we are going to map land cover using multi-seasonal Sentinel-2 imagery and a random forest classifier. The random forest classifier is an ensemble machine learning method, which uses bootstrap sampling to build many single decision tree models (Breiman 2001; Rodriguez-Galiano et al. 2012; Mellor et al. 2013). The following are the image classification steps.

Step 1: Load the necessary packages or libraries
First, load all required packages.

```
> # First, load the following libraries:
require(sp)
require(rgdal)
require(raster)
require(caret)
require(randomForest)
require(RStoolbox)
require(ggplot2)
require(reshape)
```

Step 2: Load the raster data
Set up the working directory where all data sets are located. The data sets consist of Sentinel-2 imagery (i.e., raster data in "tif" format) and training data polygons (in shapefile format).

```
> # Set up the working directory:
> setwd("/home/Sentinel/2020/Sentinel-2/")
```

Create a list of raster bands that will be used for classification.

```
> # Create a list of raster bands that will be used for classification.
> rlist=list.files(getwd(),pattern = "tif$", full.names = TRUE)
```

Combine or stack the raster layers.

```
> # Combine or stack the raster layers.
> rvars <- stack(rlist)
```

Next, check the attributes of Sentinel-2 imagery.

```
> # Check the summary statistics Sentinel-2 imagery
> rvars
class        : RasterStack
dimensions   : 4806, 5950, 28595700, 27  (nrow, ncol, ncell, nlayers)
resolution   : 10, 10  (x, y)
exten        : 260350, 319850, 8001640, 8049700  (xmin, xmax, ymin, ymax)
crs          : +proj=utm +zone=36 +south +datum=WGS84 +units=m +no_defs +ellps=WGS84 +towgs84=0,0,0
names        : S2_Apr_Jun2020.1, S2_Apr_Jun2020.2, S2_Apr_Jun2020.3, S2_Apr_Jun2020.4, S2_Apr_Jun2020.5,
S2_Apr_Jun2020.6, S2_Apr_Jun2020.7, S2_Apr_Jun2020.8, S2_Apr_Jun2020.9, S2_Jan_Mar2020.1,
S2_Jan_Mar2020.2, S2_Jan_Mar2020.3, S2_Jan_Mar2020.4, S2_Jan_Mar2020.5, S2_Jan_Mar2020.6,
S2_Jan_Mar2020.7, S2_Jan_Mar2020.8
```

Display the composite Sentinel-2 TM imagery using the ***plotRGB()*** function.

```
# Display the composite Sentinel-2 imagery in false color
S2_multiseasonal < - plotRGB(rvars, r = 7, g = 3, b = 2, stretch = "lin")
```

Step 3: Load the training data

Load the shapefile that contains the training data using the ***readOGR()*** function.

```
> # Read training data shapefile.
> ta_data < - readOGR(getwd(), "Hre_Revised_Polygon_TA_2020")
OGR data source with driver: ESRI Shapefile
Source: "/home/kamusoko/Documents/Projects/DL_ML_Apps/GEE_Classifications/Urban/
Harare/SatelliteImagery/Sentinel/2020/
Sentinel-2/Sentinel-2_All_Bands/Chapter1/Sentinel_2/Original", layer: "Hre_Revised_Polygon_TA_2020"
with 5711 features
It has 1 field
```

The training data has one field (column) with 5,711 land cover classes.
Next, use the *summary()* command to check the distribution of the land cover classes.

```
> # Check the land cover class distribution.
> summary(ta_data)
```

Object of class SpatialPolygonsDataFrame

```
      Class
Bare areas  :1091
Built-up    :2113
Cropland    :1008
Grass/open  :1095
Water       : 73
Woodlands   : 331
```

Plot the training data on the Sentinel-2 imagery (Fig. 3.3).

```
> # Plot training data
> olpar < - par(no.readonly = TRUE) # back-up par
> par(mfrow = c(1,2))
> colors < - c("grey","red","yellow","light green","blue","green")
> plot(ta_data, add = TRUE, col = colors[ta_data$Class], pch = 19)
```

Step 4: Train the random forest model

Next, set up and run the random forest (RF) model. We are going to use the *superClass()* function in the **RSToolbox** package to set up the classifier (Horning et al. 2016). First, we are going to set up a seed to ensure that we get the same result when we run the model again. We are also going to record the time for running the model.

We define the classifier parameters as follows. The *superClass()* refers to supervised classification function. The object *rvars* contain Sentinel-2 imagery (to be classified), while *trainData* is a spatial polygon dataframe that contains training areas (*ta_data*). The *responseCol* (i.e., the response variable) is a column (*Class* in this case) in *trainData*, which can be either a character or an integer. Note the *responseCol* can be omitted when *trainData* has only one column. We select the random forest as the *model* with a *tuneLength* of 3, which is the number of levels for each tuning parameter. However, we are going to use the default cross-validation, which is 5. Note this represents the number of cross-validation resamples during model tuning. Finally, 70% of the training data set will be used for training, while 30% will be used for model validation. In particular, the training set will be used to find the optimal model parameters as well as to check initial model performance based on repeated cross-validation. Note that final accuracy assessment will done in Chap. 6.

Fig. 3.3 Training data (yellow polygons) overlaid on Sentinel-2 composite false color imagery: red (nir band), green (red band), and blue (green band)

```
> ## Fit random forest model (splitting training into 70% training data, 30% validation data)

> # Set a pre-defined value using set.seed()
> hre_seed < - 27
> set.seed(hre_seed)

> # Set-up timer to check how long it takes to run the model
> timeStart < - proc.time()

> # Set-up the model parameters and run it.
> rf_SC < - superClass(rvars, trainData = ta_data, responseCol = "Class",
+    model = "rf", tuneLength = 3, trainPartition = 0.7)

> proc.time() - timeStart
    user    system   elapsed
1665.346   15.922   1681.153
```

Note that on an Intel® Core™ i7-9750H processor with a 32 GB memory, it took approximately 28 min to train the RF model with 27 Sentinel-2 bands.

Next, check the RF model performance.

```
> # Check RF model parameters
> rf_SC$model


Random Forest
7620 samples
    27 predictor
      6 classes: 'Bare areas,' 'Built-up,' 'Cropland,' 'Grass/open,' 'Water,' and 'Woodlands.'
No preprocessing
Resampling: Cross-Validated (fivefold)
Summary of sample sizes: 6096, 6095, 6097, 6095, 6097
```

```
Resampling results across tuning parameters:

mtry       Accuracy          Kappa
2          0.9066925         0.8875208
14         0.9099718         0.8914999
27         0.9062980         0.8870700
```

```
Accuracy was used to select the optimal model using the largest value.
The final value used for the model was mtry = 14.
```

The results show that 7620 training samples were used for training. The 27 predictor variables are the Sentinel-2 bands, while the six land cover classes represent the response (target) variable. The best model had an mtry value of 14 with an overall accuracy of 91%, which is relatively good (Fig. 3.4).

Next, display the RF model training performance (Fig. 3.4).

```
# Plot CV model
plot(rf_SC$model)
```

Check the parameters of the best model.

```
> # Check the parameters of the best model.
> rf_SC$model$finalModel
```

```
Call:
randomForest(x = x, y = y, mtry = param$mtry)
                Type of random forest: classification
                      Number of trees: 500
Number of variables tried at each split: 14
          OOB estimate of error rate: 8.28%
Confusion matrix:
```

	Bare areas	Built-up	Cropland	Grass/open	Water	Woodlands	class.error
Bare areas	1033	118	32	34	1	2	0.153278689
Built-up	54	1331	33	16	0	1	0.072473868
Cropland	17	39	1375	80	0	4	0.092409241
Grass/open	21	12	117	1152	0	19	0.127933384
Water	0	0	0	1	1026	1	0.001945525
Woodlands	0	1	6	21	1	1072	0.026339691

The output shows a confusion matrix for the best model (after cross-validation). A total of 500 decision trees were used in the RF model. From the 27 predictor variables (27 Sentinel-2 bands), only 14 predictor variables (bands) were selected at each split. The out-of-bag (OOB) estimate of error rate is 8.2%.

Next, display variable importance using **varImp()**function (Fig. 3.5).

```
# Compute Variable Importance
SC_varImp < - varImp(rf_SC$model, compete = FALSE)
ggplot(SC_varImp, top = 10)
```

Figure 3.5 shows the relative importance of the top 10 predictors. The top three predictors are from the post-rainy season Sentinel-2 imagery (SWIR2 and red bands), followed by Sentinel-2 blue band (S2_Jan_Mar2020.1) from the rainy season. The green (S2_Jan_Mar2020.2) and red (S2_Jan_Mar2020.3) bands from the rainy season and the SWIR1 (S2_Apr_Jun2020.8) and blue bands (S2_Apr_Jun2020.1) from the post-rainy season had variable importance between 25 and 50. In contrast, Vegetation red edge1 (S2_Jul_Sept2020.4) and SWIR1 (S2_Jul_Sept2020.1) from the dry season had variable importance less than 25. The variable importance results indicate that Sentinel-2 imagery from the post-rainy and rainy seasons has more impact in the RF model. Figure 3.2b shows that the post-rainy season SWIR2 band (S2_Apr_Jun2020.9) and Red band (S2_Apr_Jun2020.3) were better at separating land cover classes. The variable importance results suggest that land cover analysis in the study area should be performed using rainy and post-rainy season satellite imagery.

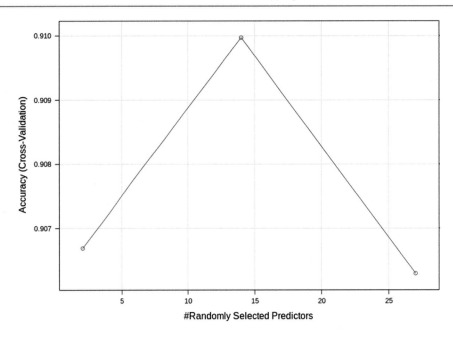

Fig. 3.4 Repeated cross-validation (based on overall accuracy) profile for the RF model

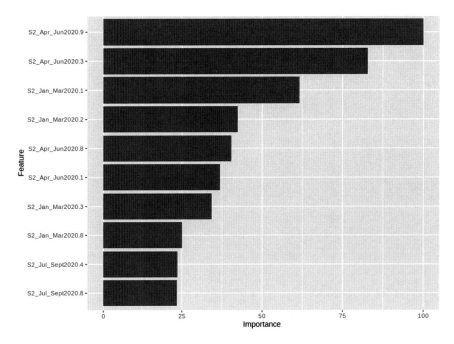

Fig. 3.5 Random forest variable importance

Next, perform validation of the RF model. First, we use the RF model results for prediction and then build a confusion matrix as shown below.

```
> # Checkmodel validation
> rf_SC

superClass results
*********** Validation **************
$validation
Confusion Matrix and Statistics
       Reference
Prediction      Bare areas    Built-up    Cropland    Grass/open    Water    Woodlands
Bare areas      720           81          38          68            0        1
Built-up        204           969         56          28            0        0
Cropland        103           22          1034        284           0        2
Grass/ open     50            8           69          712           0        20
Water           7             6           0           0             1012     8
Woodlands       1             3           4           9             0        1005


Overall Statistics
              Accuracy: 0.8357
                95% CI: (0.8265, 0.8446)
   No Information Rate: 0.1841
   P-Value [Acc>NIR]   :<2.2e-16
                 Kappa: 0.8026
 Mcnemar's Test P-Value: NA
Statistics by Class:
                     Class:      Class:      Class:      Class:       Class:     Class:
                     Bare areas  Built-up    Cropland    Grass/ open  Water      Woodlands
Sensitivity          0.6636      0.8898      0.8609      0.6467       1.0000     0.9701
Specificity          0.9654      0.9470      0.9228      0.9729       0.9962     0.9969
Pos Pred Value       0.7930      0.7709      0.7156      0.8289       0.9797     0.9834
Neg Pred Value       0.9350      0.9772      0.9671      0.9313       1.0000     0.9944
Prevalence           0.1663      0.1669      0.1841      0.1688       0.1551     0.1588
Detection Rate       0.1104      0.1485      0.1585      0.1091       0.1551     0.1540
Detection Prevalence 0.1392      0.1927      0.2215      0.1317       0.1583     0.1567
Balanced Accuracy    0.8145      0.9184      0.8919      0.8098       0.9981     0.9835
```

The confusion matrix shows that the overall classification accuracy is 83.6%, which is relatively good. There are high errors of omission for the bare areas class as most built-up (204), cropland (103), and grass/ open areas (50) pixels are excluded. The built-up areas class include misclassified bare areas and cropland pixels. However, there is a substantial increase in commission errors for the built-up class since it includes 204 misclassified bare areas pixels. This indicates spectral confusion between the built-up and bare areas. With respect to the cropland class, high commission errors are observed given that this class includes substantial misclassified grass/open areas (284) and cropland (103) pixels. For the grass/open areas, there is also a significant amount of cropland pixels that have been misclassified. This indicates great difficulty to separate cropland and grass/open areas. Generally, woodland and water have relatively high classification accuracies.

The class accuracy metrics reveal very important information, which may be used to improve the model. In this workbook, we focus on the producer's accuracy which is the "Sensitivity" metric, and the user's accuracy which is the "Pos Pred Value" because these are commonly used for accuracy assessment in remote sensing literature. The producer's accuracy (sensitivity) is significantly lower than the user's accuracy (Pos Pred Value) for the bare areas class. This indicates high omission errors and thus an underestimation of bare areas. With respect to the built-up class, the producer's accuracy (sensitivity) is higher than the user's accuracy (Pos Pred Value). This indicates that the RF model overestimated built-up areas given the high commission errors. Similarly, the producer's accuracy (sensitivity) is higher than the user's accuracy (Pos Pred Value) for the cropland class, which indicates relatively high commission errors and thus an overestimation of the

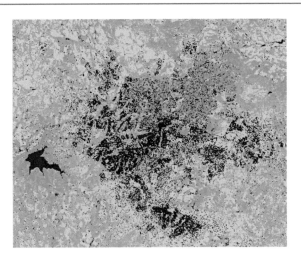

Fig. 3.6 Land cover classification based on multi-seasonal Sentinel-2 imagery and a random forest model

cropland areas. In contrast, the producer's accuracy is substantially lower than the user's accuracy for the grass/open areas class. This indicates high omission errors and thus an underestimation of the grass/open areas. Generally, the woodland and water classes have high individual accuracies, which indicates relatively good model performance for these classes. While the overall accuracy is relatively good, major classification errors are still present. For example, the RF model has difficulties to separate built-up and bare areas on one hand, and cropland and grass/open areas on the other hand.

Display the land cover map (Fig. 3.6).

```
> # Display the land cover map
> plot(rf_SC$map, col = colors, legend = FALSE, axes = FALSE, box = FALSE)
> legend(1,1, legend = levels(ta_data$Class), fill = colors, title = "Classes", horiz = TRUE, bty = "n")
```

We are going to apply a majority filter based on a 3 × 3 window filter in order to remove small pixels that cause a salt and pepper effect on the land cover map (Fig. 3.6).

```
> # Filter the land cover map.
> # Create a 3 × 3 window filter
> window < - matrix(1, 3, 3)

> # Perform majority filtering
> RF_2020_MF < - focal(rf_SC$map, w = window, fun = modal)
```

Step 5: Save the final land cover map

Finally, save the land cover map so that it can be displayed in QGIS or other GIS software.

```
# Save land cover map
writeRaster(RF_2020_MF, filename = "S2_MultiSeasonal_LC_Jan_Sep_2020a.img", type = "raw",
        datatype = 'INT1U', index = 1, na.rm = TRUE, progress = "window", overwrite = TRUE).
```

3.2.2   Lab 2. Mapping Land Cover Using Multi-seasonal Sentinel-2 Imagery and Other Derived Data

Objective

- Map land cover using multi-seasonal Sentinel-2 imagery, spectral and texture indices, and a random forest classifier.

Procedure

In this lab, we are going to map land cover using multi-seasonal Sentinel-2 imagery, spectral and texture indices, and a random forest (RF) classifier. The spectral and texture indices that were computed from Sentinel-2 imagery in Chap. 2 will be used in this lab. The following are the image classification steps.

Step 1: Load the necessary packages or libraries
First, load all required packages.

```
> # First, load the following libraries:
require(sp)
require(rgdal)
require(raster)
require(caret)
require(randomForest)
require(RStoolbox)
require(ggplot2)
require(reshape)
```

Step 2: Load the raster data
Set up the working directory as we have done in lab 1. The directory contains Sentinel-2 imagery and reference training data polygons.

```
> # Set up the working directory:
> setwd("/home/Sentinel/2020/Sentinel-2/")
```

Create a list of raster bands that will be used for classification.

```
> # Create a list of raster bands that will be used for classification.
> rlist = list.files(getwd(),pattern = "tif$", full.names = TRUE)
```

Combine or stack the raster layers, which you listed before.

```
> # Combine or stack the raster layers.
> rvars < - stack(rlist)
```

Next, check the attributes of Sentinel-2 imagery.

```
> # Check the summary statistics
> rvars
dimensions  : 4806, 5950, 28,595,700, 67 (nrow, ncol, ncell, nlayers)
resolution  : 10, 10 (x, y)
extent      : 260350, 319850, 8001640, 8049700 (xmin, xmax, ymin, ymax)
crs         : + proj = utm + zone = 36 + south + ellps = WGS84 + towgs84 = 0,0,0,0,0,0,0 + units = m + no_defs.
names       : mndwi_post_rainy_2020, ndbi_post_rainy_2020, ndbi_rainy_season_2020, ndvi_-
post_rainy_2020, osavi_rainy_season_2020, pr_entropy_S2_b2, pr_entropy_S2_b3,
```

Step 3: Load the training data

Load the shapefile that contains training data using the ***readOGR()*** function.

```
> # Read training data shapefile.
> ta_data < - readOGR(getwd(), "Hre_Revised_Polygon_TA_2020")
OGR data source with driver: ESRI Shapefile
Source: "/
home/kamusoko/Documents/Projects/DL_ML_Apps/GEE_Classifications/Urban/Harare/SatelliteImagery/Sentinel/
2020/Sentinel-2/Sentinel-2_All_Bands/Chapter1/Sentinel_2/Original", layer: "Hre_Revised_Polygon_TA_2020"
with 5711 features
It has 1 fields
```

The training data has one field (column) with 5,711 land cover classes.
Next, use the *summary()* command to check the distribution of the land cover classes.

```
> # Check the land cover class distribution.
> summary(ta_data)
Object of class SpatialPolygonsDataFrame
Coordinates:
        min           max
x   260516.9      319789.7
y   8,001,673.2   8,049,677.8
Is projected: TRUE
proj4string :
[+proj = utm + zone = 36 + south + datum = WGS84 + units = m + no_defs + ellps = WGS84 + towgs84 = 0,0,0].
Data attributes:
      Class
Bare areas   :1091
Built-up     :2113
Cropland     :1008
Grass/open   :1095
Water        : 73
Woodlands    : 331
```

Step 4: Train the random forest model

We are going to define the same parameters as we did in lab 1 in order to run the RF model within *superClass()* function from **RSToolbox** package.

```
> ## Fit random forest model (splitting training into 70% training data, 30% validation data)

> # Set a pre-defined value using set.seed()
> hre_seed < - 27
> set.seed(hre_seed)
> # Set-up timer to check how long it takes to run the model
> timeStart < - proc.time()
> # Set-up the model parameters and run it.
> rf_SC      < - superClass(rvars, trainData = ta_data, responseCol = "Class",
+                           model = "rf", tuneLength = 3, trainPartition = 0.7).
> proc.time() - timeStart
   user     system    elapsed
2157.576   39.892    2197.279
```

It took approximately about 36 min to train the RF model with the 67 predictors (Sentinel-2 bands and spectral and textural indices). Next, check the RF model performance.

```
> # Check RF model parameters
> rf_SC$model


Random Forest
7620 samples
   67 predictor
   6 classes: 'Bare areas,' 'Built-up,' 'Cropland,' 'Grass/open,' 'Water,' and 'Woodlands.'.
No pre-processing
Resampling: Cross-Validated (fivefold)
Summary of sample sizes: 6096, 6095, 6097, 6095, 6097
Resampling results across tuning parameters:


mtry        Accuracy            Kappa
2           0.9060369           0.8867050
34          0.9077421           0.8888041
67          0.9018371           0.8816850


Accuracy was used to select the optimal model using the largest value.
The final value used for the model was mtry = 34.
```

The results show that 7620 training samples were used for training. The 67 predictor variables are the Sentinel-2 bands and other derived Sentinel-2 variables (spectral and textural indices), while the six land cover classes represent the response (target) variable. The best model had an mtry value of 34 with an overall accuracy of 91%, which is relatively good (Fig. 3.7).

Next, display the RF model training performance (Fig. 3.7).

```
# Plot CV model
plot(rf_SC$model)
```

Fig. 3.7 Repeated cross-validation (based on overall accuracy) profile for the RF model

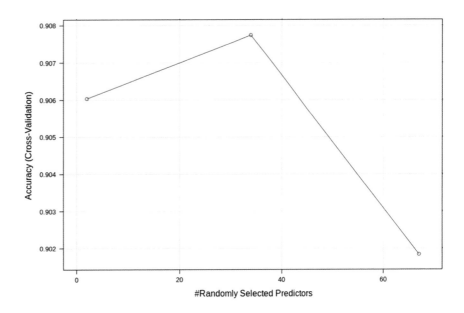

Check the parameters of the best model.

```
> # Check the parameters of the best model.
> rf_SC$model$finalModel
```

```
Call:

randomForest(x = x, y = y, mtry = param$mtry)
                        Type of random forest: classification
                              Number of trees: 500
     Number of variables tried at each split   : 34
                     OOB estimate of error rate: 8.74%
Confusion matrix:
               Bare areas    Built-up    Cropland    Grass/open    Water    Woodlands    class.error
Bare areas     1001          147         31          39            0        2            0.179508197
Built-up       46            1341        31          16            0        1            0.065505226
Cropland       12            42          1357        100           0        4            0.104290429
Grass/open     22            25          103         1155          0        16           0.125662377
Water          0             1           0           0             1026     1            0.001945525
Woodlands      1             0           6           19            1        1074         0.024523161
```

The output shows a confusion matrix for the best model (after cross-validation). A total of 500 decision trees were used in the RF model. From the 67 predictor variables, only 34 predictor variables were selected at each split. The out-of-bag (OOB) estimate of error rate is 8.7%, which slightly higher than the previous model in lab 1.

Next, display variable importance using ***varImp()***function (Fig. 3.8).

```
# Compute Variable Importance
SC_varImp < - varImp(rf_SC$model, compete = FALSE)
ggplot(SC_varImp, top = 10)
```

Figure 3.8 shows the contribution of the top 10 predictors. The top four predictors with a variable importance over 50 are from the post-rainy season Sentinel-2 SWIR2 band (S2_Apr_Jun2020.9), rainy season OSAVI, rainy season Sentinel-2 mean texture index, and the post-rainy season Sentinel-2 NIR band (S2_Apr_Jun2020.8) from the rainy season. The other six predictors from the rainy and post-rainy season have variable importance below 50. This also shows that predictors from the post-rainy and rainy seasons have more impact in the model. Furthermore, OSAVI, mean texture, NDVI, and variance texture indices have significance importance in RF classifier. This also suggests that land cover mapping in the study area should also include additional variables.

Fig. 3.8 Random forest variable importance

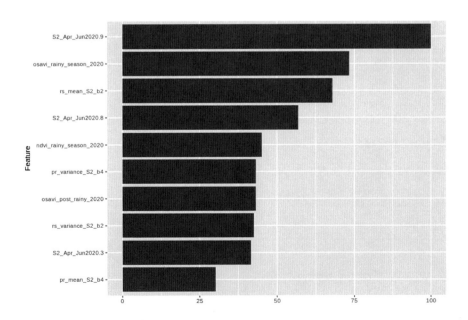

Next, perform validation of the random forest classifier. First, we use the model results for prediction and then build a confusion matrix as we have done in previous exercises.

```
> # Checkmodel validation
> rf_SC


superClass results
*********** Validation **************
$validation
Confusion Matrix and Statistics
        Reference
Prediction      Bare areas    Built-up    Cropland    Grass/open    Water    Woodlands
Bare areas      736           62          51          55            0        0
Built-up        140           993         24          16            0        0
Cropland        127           19          1053        413           0        2
Grass/open      73            7           71          606           0        14
Water           8             6           0           0             1012     0
Woodlands       1             2           2           11            0        1020


Overall Statistics
                Accuracy: 0.8308
                  95% CI: (0.8215, 0.8398)
     No Information Rate: 0.1841
     P-Value [Acc > NIR]: < 2.2e-16
                   Kappa: 0.7966
 Mcnemar's Test P-Value  : NA
Statistics by Class:
                     Class:       Class:       Class:       Class:        Class:      Class:
                     Bare areas   Built-up     Cropland     Grass/ open   Water       Woodlands
Sensitivity          0.6783       0.9118       0.8768       0.55041       1.0000      0.9846
Specificity          0.9691       0.9669       0.8946       0.96957       0.9975      0.9971
Pos Pred Value       0.8142       0.8465       0.6524       0.78599       0.9864      0.9846
Neg Pred Value       0.9379       0.9821       0.9699       0.91396       1.0000      0.9971
Prevalence           0.1663       0.1669       0.1841       0.16876       0.1551      0.1588
Detection Rate       0.1128       0.1522       0.1614       0.09289       0.1551      0.1563
Detection Prevalence 0.1386       0.1798       0.2474       0.11818       0.1573      0.1588
Balanced Accuracy    0.8237       0.9394       0.8857       0.75999       0.9987      0.9908
```

The confusion matrix shows that the overall classification accuracy is 83.1%, which is slightly lower than the previous results in lab 1 (83.6%). The correctly classified metric significantly improved for all classes except grass/open areas class. However, the omission errors for the bare areas class are still high as most built-up (140), cropland (127), and grass/open areas (73) pixels are excluded. For the built-up areas class, a slight decrease in commission errors is observed compared to the previous results in lab 1. However, there are high commission errors for built-up areas class since it includes 140 bare areas pixels. With regard to the cropland class, high omission errors are observed as most bare areas (127) and grass/open areas (413) pixels are falsely included in this class. Clearly, there is spectral confusion between cropland and bare areas on the one hand and between cropland and grass/open areas on the other hand. For the grass/open areas, there is also an increased amount of cropland pixels that have been misclassified. It is noteworthy that the RF model had difficulty to separate cropland and grass/open areas. Generally, woodland and water have relatively high accuracies. This indicates that spectral confusion is still problem despite the fact we are using multi-seasonal imagery (from the rainy, post-rainy, and dry season) plus additional spectral and texture indices derived from Sentinel-2 imagery.

The class accuracy metrics also reveal very important insights. As was observed in the previous lab, the producer's accuracy (sensitivity) is significantly lower than the user's accuracy (Pos Pred Value) for the bare areas. This indicates high omission errors and an underestimation of the bare areas class. Although the built-up areas class has high individual accuracies, the producer's accuracy (sensitivity) is relatively higher than the user's accuracy (Pos Pred Value). This indicates an overestimation of built-up areas. With respect to the cropland class, the producer's accuracy (sensitivity) is significantly higher than the user's accuracy (Pos Pred Value) indicating high commission errors and thus an overestimation of this class. For the grass/open areas class, the producer's accuracy is substantially lower than the user's accuracy indicating high omission errors and thus an underestimation of this class. This is attributed to spectral confusion between grass/open areas and cropland class. Generally, the woodland and water classes have high individual accuracies. It is noteworthy that class accuracies for built-up areas increased substantially. This suggests additional spectral and texture indices derived from multi-seasonal Sentinel-2 data and spectral and texture indices improved the classification of built-areas. However, the RF classifier is having difficulty in separating cropland and grass/open areas classes in the study area. This is because cropland and grass/open areas are spectrally similar during the post and dry seasons when croplands are covered with grass cover (Fig. 3.2). Therefore, NDVI and OSAVI are not adequate to separate cropland and grass/open areas classes. An alternative is to check if spectral index that incorporates the vegetation red edge (2 and 3) and NIR could minimize spectral confusion between cropland and grass/open areas.

Display the land cover map (Fig. 3.9).

```
> # Display the land cover map
> colors < - c("grey","red", "yellow", "light green", "blue", "green")
> plot(rf_SC$map, col = colors, legend = TRUE, axes = FALSE, box = FALSE)
> legend(1,1, legend = levels(ta_data$Class), fill = colors, title = "Classes", horiz
      = TRUE, bty = "n")
```

We are going to apply a majority filter based on a 3 × 3 window filter in order to remove small pixels that cause a salt and pepper effect on the land cover map (Fig. 3.9).

```
> # Filter the land cover map.
> # Create a 3 × 3 window filter
> window < - matrix(1, 3, 3)

> # Perform majority filtering
> RF_2020_MF < - focal(rf_SC$map, w = window, fun = modal)
```

Step 5: Save the final land cover map
Finally, save the land cover map so that it can be displayed in QGIS or other GIS software.

```
# Save land cover map
writeRaster(RF_2020_MF, filename = "S2_MS_SV_Tex_LC_Jan_Sep_2020b.img", type = "raw",
      datatype = 'INT1U', index = 1, na.rm = TRUE, progress = "window", overwrite = TRUE).
```

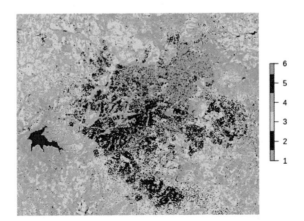

Fig. 3.9 Land cover classification based on multi-seasonal Sentinel-2 imagery, spectral and texture indices, and a random forest classifier

3.3 Summary

In this chapter, we completed two land cover mapping labs using multi-seasonal Sentinel-2 and spectral and texture indices. In the first lab, we used only multi-seasonal Sentinel-2 imagery, while in the second lab, multi-seasonal Sentinel-2 and spectral and texture indices were used. The land cover map produced in the first lab had a slightly higher accuracy than the land cover map produced in lab 2 (83.6% vs. 83.1%). While the overall classification accuracy did not increase in lab 2—as is usually reported in literature—a significant improvement was observed for the built-up class. This suggests that additional Sentinel-2 derived spectral and textural indices used in lab 2 had an impact in improving the accuracy of built-up class. However, spectral confusion is still observed. This suggests that additional satellite data or variables are required to minimize spectral confusion and therefore improve land cover classification accuracy.

3.4 Additional Exercises

i. Produce land cover maps using multi-seasonal Sentinel-2 imagery, spectral and texture indices, and support vector machines (SVM).
ii. Compare SVM and RF classification results.
iii. Discuss the performance of the SVM and RF classifiers in terms of model performance and validation.

References

Breiman L (2001) Random forests. Mach Learn 45:5–32

Griffiths P, Hostert P, Gruebner O, van der Linden S (2010) Mapping megacity growth with multi-sensor data. Remote Sens Environ 114:426–439

Herold M, Roberts DA, Gardner ME, Dennison PE (2004) Spectrometry for urban area remote sensing-development and analysis of a spectral library from 350 to 2400 nm. Remote Sens Environ 91:304–319

Horning N, Leutner B, Wegmann M (2016) Land cover or image classification approaches. In Wegmann M, Leutner B, Dech S (eds) Remote sensing and GIS for ecologists: using Open Source Software. Exeter: Pelagic Publishing, UK

Jensen JR (2000) Remote sensing of the environment: an Earth resource perspective. Pearson Education, pp 407–470

Kamusoko C, Gamba J, Murakami H (2013) Monitoring urban spatial growth in Harare Metropolitan province, Zimbabwe. Adv Remote Sens 2:322–331

Kamusoko C, Kamusoko OW, Chikati E, Gamba J (2021) Mapping urban and peri-urban land cover in Zimbabwe: challenges and prospects. Geomatics 1:114–147

Korn J, Rabe A, Radeloff VC, Kuemmerle T, Kozak J, Hostert P (2009) Land cover mapping of large areas using chain classification of neighboring Landsat satellite images. Remote Sens Environ 113(5):957–964

Mellor A, Haywood A, Stone C, Jones S (2013) The performance of random forests in an operational setting for large area sclerophyll forest classification. Remote Sens 5(6):2838–2856

Rodriguez-Galiano VF, Chica-Olmo M, Abarca-Hernandez F, Atkinson PM, Jeganathan C (2012) Random Forest classification of Mediterranean land cover using multi-seasonal imagery and multi-seasonal texture. Remote Sens Environ 121:93–107

Schneider A (2012) Monitoring land cover change in urban and peri-urban areas using dense time stacks of Landsat satellite data and a data mining approach. Remote Sens Environ 124:689–704

Schug F, Okujeni A, Hauer J, Hostert P, Nielsen JØ, van der Linden S (2018) Mapping patterns of urban development in Ouagadougou, Burkina Faso, using machine learning regression modeling with bi-seasonal Landsat time series. Remote Sens Environ 210:217–228

Small C (2004) Landsat ETM+ spectral mixing space. Remote Sens Environ 93:1–17

Wania A, Kemper T, Tiede D, Zeil P (2014) Mapping recent built-up area changes in the city of Harare with high resolution satellite imagery. Appl Geogr 46:35–44

Yuan F, Sawaya KE, Loeffelholz BC, Bauer ME (2005) Land cover classification and change analysis of the Twin Cities (Minnesota) Metropolitan area by multitemporal Landsat remote sensing. Remote Sens Environ 98:317–328

Mapping Urban Land Cover Using Multi-Seasonal Sentinel-1 Imagery and Texture Indices

4

Abstract

The previous chapter showed limitations of using optical satellite data to map land cover in urban areas. Therefore, it is important to test the effectiveness of synthetic aperture radar (SAR) data for mapping urban land cover. In this chapter, we are going to map land cover using multi-seasonal Sentinel-1 imagery, textural indices (derived from multi-seasonal Sentinel-1 imagery) and a random forest classifier.

Keywords

Land cover mapping • Synthetic aperture radar (SAR) • Multi-seasonal • Sentinel-1 • Texture indices • Random forest classifier

4.1 Introduction

4.1.1 Background

Substantial progress has been made in remotely sensed image classification due to developments in Earth Observation (EO) sensor technology coupled with rapid advancement in computer hardware and machine learning algorithms. With the advent of the Earth Resources Satellite in 1970s, EO data has been collected in various spectral, spatial, and temporal resolutions in order to map urban land cover (Goldblatt et al. 2018). To date, many studies have used Landsat data to map land cover and analyze land cover change in urban areas (Scheineder 2012; Seto et al. 2011). However, land cover mapping challenges still remain, particularly in newly developing and sparse urban areas (Zhu et al. 2012; Schug et al. 2018). This is attributed to the difficulty of detecting complex urban spatial patterns at the low and medium spatial resolutions (Lu and Weng 2007; Griffiths et al. 2010). However, the launch of the European Space Agency's Copernicus program Sentinel satellites in 2014 represents opportunities to advance land cover mapping in urban areas. Sentinel satellites provide global coverage of synthetic aperture radar (SAR) and optical data for mapping and monitoring land cover at high spatial and temporal resolutions (ESA 2015).

The past decades have witnessed an increase in the availability of SAR sensors. However, literature review shows that optical satellite sensors have been widely used to map land cover in urban areas. This is because processing and analyzing SAR data are complex compared to optical satellite data. Furthermore, SAR data has been limited for urban mapping due to their sensitivity to clustered and irregular urban structures (Dell'Acqua and Gamba 2003). For example, the so-called "cardinal effect" (Hardaway et al. 1982; Bryan 1979) produces strong backscatter from built-up areas that are oriented orthogonal to the SAR sensor. However, built-up areas that are not oriented orthogonal to the SAR sensor look direction are not captured in the SAR imagery. This is because the incident SAR beam is reflected away from the sensor. Nonetheless, some studies have reported that SAR texture imagery improves land cover mapping (Dell'Acqua and Gamba 2003). In addition, SAR data yields physical quantitative measures that can be used to map forest structure (Flores et al. 2019). Therefore, SAR data offers great potential to map land cover. In order to use SAR data efficiently, there is need to understand the signal variation and its relationship to the SAR sensor properties as well as the target area. In this chapter, we are going to briefly cover SAR basics. Note that a detailed and advanced explanation of SAR is beyond the scope of the workbook. Readers who need more detailed explanations on SAR are encouraged to consult SAR literature.

4.1.2 Synthetic Aperture Radar (SAR) Basics

Synthetic Aperture Radar (SAR) is a RAdio Detection And Ranging (radar) system that uses electromagnetic radiation (EMR) in the microwave wavelength between 3 cm and 1 m (Tso and Mather 2001). Synthetic aperture refers to the virtual creation of a large antenna based on sensor motion, which is required to ensure an adequate resolution in azimuth (Jensen 2000). SAR is an active sensor that transmits EMR and receives backscatter to the sensor (Roberts and Robertson 2016). The backscattered signal is sampled in time bins along the along-track direction of the sensor antenna (azimuth) and across-track or perpendicular to the direction of the sensor antenna (range) (Sabins 1997; Mather and Koch 2011). SAR imagery is generated from the intensity (or amplitude) and timing of the backscatter (Tso and Mather 2001). SAR sensors can transmit EMR and receive backscatter in different polarizations. Polarization refers to the orientation of the EMR that occurs in one plane. The orientation of the EMR can be horizontal or vertical to the direction of transmission (Roberts and Robertson 2016). Co-polarization VV refers to vertical transmit and vertical receive, while co-polarization HH refers to horizontal transmit and horizontal receive (Eckardt et al. 2019). Cross-polarization VH refers to vertical transmit and horizontal receive, while cross-polarization HV refers to horizontal transmit and vertical receive (Eckardt et al. 2019). Generally, polarization can greatly impact image quality and mapping capability because surfaces also modify the polarization of backscattered EMR.

In order to interpret SAR data effectively, it is important to point some key differences between SAR and optical sensors. First, SAR imagery is obtained using sensors and principles that are different from optical imagery (Jensen 2000). SAR imagery is derived from an active sensor that transmits microwave radiation with a precise set of wavelengths, while optical imagery is derived from passive optical sensors that record reflected solar radiation in the visible and infrared wavelength (Sabins 1997; Jensen 2000). SAR sensors record electromagnetic energy as a function of time rather than angular distance (Mather and Koch 2011). In general, time is precisely measured and recorded than angular distance (Sabins 1997). Second, SAR is a side-looking sensor (e.g., Sentinel-1 is right-looking), while optical sensors such as Landsat are nadir looking (Sabins 1997). Third, SAR radiation is coherent. That is, two waves of EMR are coherent when they are in phase (vibrate at the same time). In contrast, reflected sunlight is not coherent as it has a wide range of wavelengths and random phases (Sabins 1997; Jensen 2000).

Remotely sensed data derived from SAR sensors is useful for land cover mapping given its many advantages (Corbane et al. 2008). First, SAR sensors operate independently of lighting and weather conditions. Second, atmospheric absorption and scattering is negligible except at the very short radar wavelengths (Sabins 1997). Third, SAR backscatter depends on the interaction of many complex influences such as the electrical and geometric properties of the target (Hardaway et al. 1982). This means that SAR imagery captures the structure and dielectric properties of the Earth surface. This information can be used to improve land cover classification (Zhu et al. 2012). Fourth, SAR data can be calibrated, without the need for atmospheric correction, which produces consistent time series (Sabins 1997).

SAR sensors have also a number of limitations. First, SAR data is affected by geometric distortions such as layover, foreshortening, and shadow (Tso and Mather 2001). The surface topography and the orientation of slopes and aspects of the observed surface generally affect backscatter. Layover occurs when the top of an illuminated target is seen by the SAR sensor as the bottom or vice versa, while foreshortening occurs when features captured on the SAR image are perceived to have less depth or distance than in reality (Sabins 1997; Jensen 2000). Shadows occur on slopes steeper than the local incidence angle and directed away from the SAR sensor (Jensen 2000). As a result, the SAR image appears black since there is no information. Second, grainy or "salt-and-pepper" effects known as speckle is a problem (Tso and Mather 2001). Speckle is a spatially random and multiplicative noise due to the superposition of multiple backscatter sources within a SAR pixel (Flores et al. 2019). While there many techniques to reduce speckle, it is difficult to eliminate. Furthermore, speckle reduction techniques decrease SAR spatial resolution (Tso and Mather 2001). Third, SAR backscatter depends on the angle of the incident microwave radiation. Since side-looking SAR operates over a range of incidence angles along the swath, the same target will appear different depending on whether it is in near range (low incidence angle) or far range (higher incidence angle) of the swath (Tso and Mather 2001; Mather and Koch 2011). Fourth, backscatter saturation also occurs, especially in mature forests with complex stand structures (Flores et al. 2019).

4.2 Land Cover Mapping Labs

In this chapter, we are going to use analysis-ready multi-seasonal Sentinel-1 imagery composites that were compiled in Google Earth Engine (GEE). Sentinel-1 is a polar-orbiting radar imaging mission that comprises a constellation of Sentinel-1A and Sentinel-1B satellites, which provides 12-day (ground track) repeat cycle for one satellite and 6-day (ground track) repeat for two satellites (ESA 2015). According to ESA (2015), Sentinel-1, constellation provides C-band (5.6 cm) SAR acquired in different modes. The multi-seasonal Sentinel-1 imagery used in this chapter consists of mean and median rainy season Sentinel-1, mean and median post-rainy season Sentinel-1, and mean and median dry season Sentinel-1 composites. We used mean and median seasonal Sentinel-1 imagery because the imagery shows lower speckle than single-date imagery. As a result, we did not perform speckle reduction because it generally reduces spatial resolution (Kamusoko et al. 2021). Note that analysis-ready Sentinel-1 data was processed and terrain corrected in GEE.

4.2.1 Lab 1. Mapping Land Cover Using Multi-Seasonal Sentinel-1 Imagery

Objective

- Map land cover using multi-seasonal Sentinel-1 SAR imagery.

Procedure

In this lab, we are going to use mean and median VV and VH Sentinel-1 SAR imagery, which were acquired in ascending orbit.

Step 1: Load the necessary packages or libraries
First, load all required packages.

```
># First, load the following libraries.
require(sp).
require(rgdal).
require(raster).
require(caret).
require(randomForest).
require(RStoolbox).
require(ggplot2).
require(reshape).
```

Step 2: Load the raster data
Set up the working directory where Sentinel-1 imagery and training data are located.

```
># Set up the working directory:
>setwd("/home/Sentinel/2020/Sentinel-1/").
```

Create a list of raster bands that will be used for classification.

```
># Create a list of raster bands that will be used for classification.
>rlist=list.files(getwd(),pattern="tif$", full.names=TRUE).
```

Combine or stack the raster layers, which you listed before.

```
># Combine or stack the raster layers.
>rvars<- stack(rlist).
```

Next, check the attributes and summary statistics of Sentinel-1 imagery.

```
> # Check the attributes of Sentinel-1 imagery.
> rvars
class       : RasterStack
dimensions  : 4806, 5950, 28595700, 15 (nrow, ncol, ncell, nlayers)
resolution  : 10, 10 (x, y)
extent      : 260350, 319850, 8001640, 8049700 (xmin, xmax, ymin, ymax)
crs         : +proj=utm +zone=36 +south +datum=WGS84 +units=m +no_defs +ellps=WGS84 +towgs84=0,0,0
names       : Hre_mean_//r_Jun_2020, Hre_mean_//n_Mar_2020, Hre_mean_//Jul_Sep_20, Hre_S1_me//Apr_Jun_20,
> # Check the summary statistics of Sentinel-1 imagery.
> summary(rvars)
```

	Hre_S1_mean_Asc_VH_Apr_Jun_20	Hre_S1_mean_Asc_VH_Jan_Mar_20	Hre_S1_mean_Asc_VH_Jul_Sep_20	Hre_S1_mean_Asc_VV_Apr_Jun_20
Min.	-33.022491	-32.095030	-31.731482	-26.26695
1st Qu.	-21.279148	-18.555556	-23.778633	-14.23895
Median	-18.997342	-17.138398	-20.934628	-12.44298
3rd Qu.	-17.023527	-15.922067	-17.869162	-10.71627
Max.	5.942189	6.300888	5.953668	15.68684
NA's	0.000000	0.000000	0.000000	0.00000

	Hre_S1_mean_Asc_VV_Jan_Mar_20	Hre_S1_mean_Asc_VV_Jul_Sep_20	Hre_S1_median_Asc_VH_Apr_Jun_20	Hre_S1_median_Asc_VH_Jan_Mar_20
Min.	-26.65277	-26.90292	-33.302760	-30.329498
1st Qu.	-12.33072	-15.32047	-21.131863	-18.428943
Median	-10.92122	-13.30935	-18.826564	-16.961433
3rd Qu.	-9.66841	-11.05843	-16.850502	-15.725245
Max.	16.42952	16.22809	5.989729	6.544743
NA's	0.00000	0.00000	0.000000	0.000000

	Hre_S1_median_Asc_VH_Jul_Sep_20	Hre_S1_median_Asc_VV_Apr_Jun_20	Hre_S1_median_Asc_VV_Jan_Mar_20	Hre_S1_median_Asc_VV_Jul_Sep_20
Min.	-31.150546	-27.94604	-26.322618	-28.58115
1st Qu.	-23.725158	-14.24522	-12.344429	-15.26421
Median	-20.735265	-12.38868	-10.874688	-13.24225
3rd Qu.	-17.722795	-10.61609	-9.564712	-10.96076
Max.	6.067399	15.64888	16.959016	16.57761
NA's	0.000000	0.00000	0.000000	0.00000

Display the mean and median Sentinel-1 imagery (Fig. 4.1).

```
># Plot Sentinel-1 imagery (mean and median).
plot(rvars).
```

Fig. 4.1 **a** Multi-seasonal (mean and median) Sentinel-1 imagery (VV and VH polarization modes), **b** Rainy season mean Sentinel-1 imagery in false color VV (red), VH (green), and VV (blue)

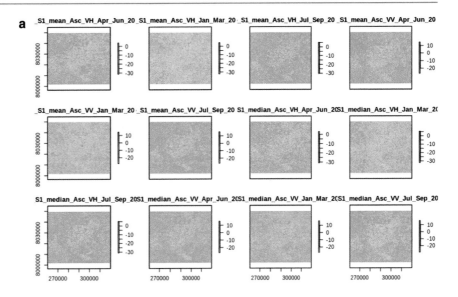

Plot the mean rainy season Sentinel-1 imagery in false color VV (red), VH (green), and VV (blue) for visualization purposes only (Fig. 4.1b).

```
># Plot Sentinel-1 imagery in false color RGB: VV (red), VH (green) and VV (blue).
>plotRGB(rvars, r="Hre_S1_mean_Asc_VV_Jan_Mar_20",g="Hre_S1_mean_Asc_VH_Jan_Mar_20",
b= "Hre_S1_mean_Asc_VV_Jan_Mar_20", stretch="lin").
```

Step 3: Load the training data
Load the shapefile that contains the training data using the ***readOGR()*** function.

```
># Read training data shapefile.
>ta_data<- readOGR(getwd(), "Hre_Revised_Polygon_TA_2020").
OGR data source with driver: ESRI Shapefile.
Source:   "/home/kamusoko/Documents/Projects/DL_ML_Apps/GEE_Classifications/Urban/Harare/SatelliteImagery/
Sentinel/2020/Sentinel-2/Sentinel-2_All_Bands/Chapter1/Sentinel_2/Original",   layer:   "Hre_Revised_Poly-
gon_TA_2020".
with 5711 features.
It has 1 field.
```

The training data has one field (column) with 5,711 land cover classes. Next, use the *summary()* command to check the distribution of the land cover classes.

Fig. 4.2 Training data (yellow polygons) overlaid on Sentinel-1 composite false color imagery: red (VV), green (VH), and blue (VV)

```
> # Check the land cover class distribution.
> summary(ta_data)
Object of class SpatialPolygonsDataFrame
 Class
 Bare areas :1091
 Built-up   :2113
 Cropland   :1008
 Grass/ open:1095
 Water      : 73
 Woodlands  : 331
```

Plot the training data on the Sentinel-1 imagery (Fig. 4.2).

```
># Plot training data.
>olpar<- par(no.readonly=TRUE) # back-up par.
>par(mfrow=c(1,2)).
>colors<- c("grey","red", "yellow", "light green", "blue", "green").
>plot(ta_data, add= TRUE, col = colors[ta_data$Class], pch =19).
```

Step 4: Train the random forest model.

Next, set up and run the random forest (RF) model. We are going to use the *superClass()* function in the **RSToolbox** package to set up the classifier.

```
>## Fit random forest classifier (splitting training into 70% training data, 30% validation data).

># Set a pre-defined value using set.seed().
>hre_seed<- 27.
>set.seed(hre_seed).

># Set-up timer to check how long it takes to run the model.
>timeStart< - proc.time().
 ># Set-up the model parameters and run it.
>rf_SC<- superClass(rvars, trainData=ta_data, responseCol="Class",
         + model = "rf", tuneLength=3, trainPartition=0.7).
>proc.time() - timeStart.
```

```
user          system        elapsed
1458.910      6.457         1465.227.
```

It took approximately 24 min to train the RF model with 12 predictors.
Check the RF model performance.

```
># Check RF model parameters.
>rf_SC$model.


Random Forest
7620 samples
  12 predictor
   6 classes: 'Bare areas', 'Built-up', 'Cropland', 'Grass/ open', 'Water', 'Woodlands'
No pre-processing
Resampling: Cross-Validated (5 fold)
Summary of sample sizes: 6096, 6095, 6097, 6095, 6097
Resampling results across tuning parameters:
```

mtry	Accuracy	Kappa
2	0.7044609	0.6441558
7	0.7041983	0.6437336
12	0.7027555	0.6420123

```
Accuracy was used to select the optimal model using the largest value.
The final value used for the model was mtry = 2.
```

The results show that 7,620 training samples were used for training. The 12 predictors represents Sentinel-1 mean and median VV and VH features, while the six land cover classes represent the response (target) variable. The best model had an mtry value of 2 with an overall accuracy of 70.5% (Fig. 4.3).

Next, display the RF model training performance (Fig. 4.3).

Fig. 4.3 Repeated cross-validation (based on overall accuracy) profile for the RF model

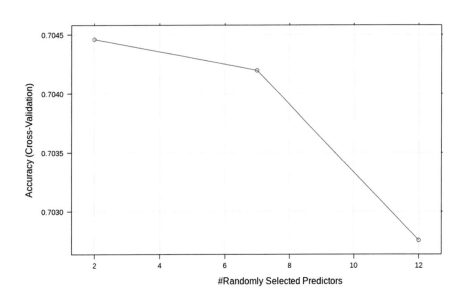

```
plot(rf_SC$model) # Plot CV model.
```

Check the parameters of the best model.

```
> # Check the parameters of the best model.
> rf_SC$model$finalModel
Call:
 randomForest(x = x, y = y, mtry = param$mtry)
               Type of random forest: classification
                     Number of trees: 500
No. of variables tried at each split: 2
        OOB estimate of  error rate: 29.33%
Confusion matrix:
```

	Bare areas	Built-up	Cropland	Grass/ open	Water	Woodlands	Class error
Bare areas	708	204	61	166	14	67	0.41967213
Built-up	76	863	50	134	0	312	0.39860627
Cropland	63	52	1148	241	0	11	0.24224422
Grass/ open	99	94	278	802	0	48	0.39288418
Water	6	2	2	1	1016	1	0.01167315
Woodlands	10	182	3	58	0	848	0.22979110

The output shows a confusion matrix for the best model (after cross-validation). A total of 500 decision trees were used in the RF model. From the 12 predictors (Sentinel-1 derived variables), only 2 predictor variables were selected at each split. The out-of-bag (OOB) estimate of error rate is 29.3%, which is worse than the OOB error in Chap. 3.

Display variable importance using ***varImp()***function (Fig. 4.4).

```
# Compute Variable Importance.
SC_varImp<- varImp(rf_SC$model, compete=FALSE).
ggplot(SC_varImp, top=10).
```

Figure 4.4 shows the relative importance of the top 10 predictors. The top four predictors are from the rainy season and dry seasons. These predictors are mean VH (Hre_S1_mean_Asc_VH_Jan_Mar_20), median VV (S1_median_Asc_VH_Jan_Mar_20), mean VV (S1_mean_Asc_VV_Jan_Mar_20), and mean VV (Hre_S1_mean_Asc_VV_Jul_Sep_20). Only two predictors are from the post-rainy season.

Fig. 4.4 Random forest variable importance

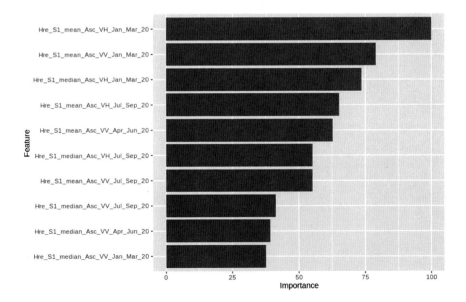

Next, perform validation of the RF model. First, we use the RF model results for prediction and then build a confusion matrix as shown in the commands below.

```
> # Checkmodel validation
> rf_SC
superClass results
************ Validation **************
$validation
Confusion Matrix and Statistics
           Reference
```

Prediction	Bare areas	Built-up	Cropland	Grass/ open	Water	Woodlands
Bare areas	587	42	56	115	12	17
Built-up	211	641	31	45	0	171
Cropland	75	5	869	253	0	2
Grass/ open	125	74	233	650	0	113
Water	10	0	0	0	1000	0
Woodlands	77	327	12	38	0	733

```
Overall Statistics
Accuracy            : 0.6867
95% CI              : (0.6753, 0.6979)
No Information Rate : 0.1841
P-Value [Acc > NIR] : < 2.2e-16
Kappa : 0.6239
```

Statistics by Class:

	Class: Bare areas:	Class: Built-up Class:	Cropland Class:	Grass/ open Class:	Water Class:	Woodlands
Sensitivity	0.54101	0.58861	0.7236	0.59037	0.9881	0.7075
Specificity	0.95551	0.91573	0.9371	0.89950	0.9982	0.9173
Pos Pred Value	0.70808	0.58326	0.7218	0.54393	0.9901	0.6175
Neg Pred Value	0.91255	0.91742	0.9376	0.91537	0.9978	0.9432
Prevalence	0.16631	0.16692	0.1841	0.16876	0.1551	0.1588
Detection Rate	0.08998	0.09825	0.1332	0.09963	0.1533	0.1124
Detection Prevalence	0.12707	0.16845	0.1845	0.18317	0.1548	0.1819
Balanced Accuracy	0.74826	0.75217	0.8303	0.74494	0.9932	0.8124

The confusion matrix shows that the overall classification accuracy is 68.7%, which is significantly lower than Sentinel-2 derived land cover classification in Chap. 3. There are high omission errors for the bare areas class since the RF model excluded built-up (211), grass/open areas (125), woodlands (77), and cropland (77) pixels. For the built-up class, there are both high commission and omission errors. For example, built-up areas include 211 misclassified bare areas pixels and 327 excluded woodland pixels. This indicates that the random forest model had difficulty to separate built-up and bare areas on one hand and built-up and woodland areas on the other hand. However, there is a significant decrease in omission and commission errors for cropland areas. In particular, there is relatively low confusion between the built-up and cropland areas. This is very significant because spectral confusion between cropland and built-up areas in optical imagery is usually one of the main causes of poor classification accuracy.

The class accuracy metrics reveal very important information, which may be used to improve the RF model. The producer's accuracy (sensitivity) is significantly lower than the user's accuracy (Pos Pred Value) for the bare areas. This indicates high omission errors, and thus, the RF model significantly underestimated the bare areas. The relatively low producer's accuracy (sensitivity) and user's accuracy (Pos Pred Value) for the built-up class indicates severe misclassification of built-up areas. In contrast, the cropland class has both high producer's and user's accuracies, and hence low omission and commission errors. For the grass/open areas class, the producer's accuracy is relatively higher than the user's accuracy indicating high commission errors. With respect to the woodland class, the producer's accuracy is higher than the user's accuracy. This indicates high commission errors and therefore an overestimation of woodland areas. Generally, the water class has highest individual accuracies, which indicates that the RF model produced an optimal classification.

Notes. Analysis of the land cover accuracy results reveals four important insights. First, built-up areas are underestimated. This is because built-up areas that are not oriented orthogonal to the S-1 sensor look direction are not captured in the S-1 imagery since the incident SAR beam is reflected away from the sensor. Second, some built-up areas in core developed urban settlements that are not oriented orthogonal to the S-1 sensor look direction are misclassified as woodland areas. However, woodlands on hills or mountain slopes that are oriented in the north-east to south-west direction are misclassified as built-up areas. Third, built-up areas in developing peri-urban and developed core urban areas that are oriented from the north-west to south-east direction are correctly classified. This is due to the double bounce and "cardinal" effects, which occurs when man-made structures are orthogonal to the SAR illumination direction (Jensen 2000). As a result, the built-up areas appear brighter due to a strong SAR backscatter. Fourth, cropland areas are not confused with built-up areas. This is very significant because spectral confusion between cropland and built-up areas in optical imagery is usually one of the main causes of poor classification accuracy.

Display the land cover map. The land cover map show poor accuracy, especially for the built-up class (Fig. 4.5).

```
># Display the land cover map.
>plot(rf_SC$map, col=colors, legend=FALSE, axes=FALSE, box=FALSE).
>legend(1,1, legend=levels(ta_data$Class), fill =colors, title= "Classes", horiz= TRUE, bty="n").
```

We are going to apply a majority filter based on a 3 × 3 window filter in order to remove small pixels that cause a salt-and-pepper effect on the land cover map (Fig. 4.5).

```
># Filter the land cover map.
># Create a 3×3 window filter.
>window<- matrix(1, 3, 3).
```

Fig. 4.5 Land cover classification based on multi-seasonal Sentinel-1 imagery and a random forest classifier

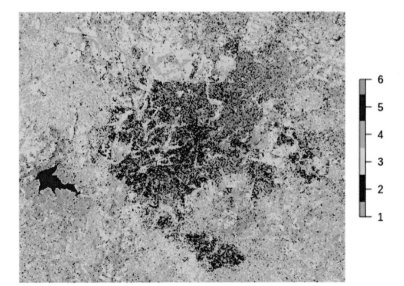

```
>#Perform majority filtering.
>RF_2020_MF<- focal(rf_SC$map, w=window, fun=modal) .
```

Step 5: Save the final land cover map.
Finally, save the land cover map so that it can be displayed in QGIS or other GIS software.

```
# Save land cover map.
writeRaster(RF_2020_MF, filename="S1_MultiSeasonal_LC_Jan_Sep_2020a.img", type="raw",
          datatype='INT1U',index=1, na.rm=TRUE, progress="window", overwrite=TRUE).
```

4.2.2 Lab 2. Mapping Land Cover using Multi-seasonal Sentinel-1 Imagery and Texture Indices

Objective

- Map land cover using multi-seasonal Sentinel-1 SAR imagery and texture indices.

Procedure
In this lab, we are going to use VV and VH Sentinel-1 SAR imagery and texture indices derived from Sentinel-1 SAR imagery.

Step 1: Load the necessary packages or libraries.
First, load all required packages.

```
>#First, load the following libraries:
require(sp).
require(rgdal).
require(raster).
require(caret).
require(randomForest).
require(RStoolbox).
require(ggplot2).
require(reshape).
```

Step 2: Load the raster data.
Set up the working directory where all data sets are located.

```
> # Set up the working directory:
> setwd("/home/Sentinel/2020/Sentinel-1/").
```

Create a list of raster bands that will be used for classification.

```
> # Create a list of raster bands that will be used for classification.
> rlist=list.files(getwd(),pattern="tif$", full.names=TRUE).
```

Combine or stack the raster layers, which you listed before.

```
> # Combine or stack the raster layers.
> rvars< - stack(rlist).
```

Next, check the attributes of Sentinel-1 imagery.

```
> # Check the attributes of Sentinel-1 imagery.
> rvars
class       : RasterStack
dimensions  : 4806, 5950, 28595700, 27 (nrow, ncol, ncell, nlayers)
resolution  : 10, 10 (x, y)
extent      : 260350, 319850, 8001640, 8049700 (xmin, xmax, ymin, ymax)
crs         : +proj=utm +zone=36 +south +ellps=WGS84 +towgs84=0,0,0,0,0,0,0 +units=m +no_defs
names       : entropy_S1_VH_rainy, entropy_S1_VV_rainy, entropy_S1_VV_VH_rainy, homogeneity_S1_VH_rainy,
homogeneity_S1_VV_rainy, homogeneity_S1_VV_VH_rainy, Hre_mean
```

Step 3: Load the training data.
Load the shapefile that contains the training data using the ***readOGR()*** function.

```
> # Read training data shapefile.
> ta_data< - readOGR(getwd(), "Hre_Revised_Polygon_TA_2020").
OGR data source with driver: ESRI Shapefile.
Source:   "/home/kamusoko/Documents/Projects/DL_ML_Apps/GEE_Classifications/Urban/Harare/SatelliteImagery/
Sentinel/2020/Sentinel-2/Sentinel-2_All_Bands/Chapter1/Sentinel_2/Original", layer: "Hre_Revised_Polygon_
TA_2020".
with 5711 features.
It has 1 field.
```

The training data has one field (column) with 5711 land cover classes. Next, use the *summary()* command to check the distribution of the land cover classes.

```
> # Check the land cover class distribution.
> summary(ta_data)
Object of class SpatialPolygonsDataFrame
 Class
 Bare areas  :1091
 Built-up    :2113
 Cropland    :1008
 Grass/ open:1095
 Water       : 73
 Woodlands   : 331
```

Step 4: Train the random forest model.
Next, set up and run the RF model. We are going to use the *superClass()* function in the **RSToolbox** package to set up the classifier.

```
> ## Fit random forest classifier (splitting training into 70% training data, 30% validation data).

> # Set a pre-defined value using set.seed().
> hre_seed<- 27.
> set.seed(hre_seed).

> # Set-up timer to check how long it takes to run the model.
> timeStart <- proc.time().
> # Set-up the model parameters and run it.
> rf_SC<- superClass(rvars, trainData=ta_data, responseCol="Class",
                + model="rf", tuneLength=3, trainPartition=0.7).
```

```
> proc.time() - timeStart.
      User system elapsed.
1420.958 10.760 1431.698.
```

It took approximately 24 min to train the RF model with 20 predictors.
Check the RF model performance.

```
> # Check RF model parameters.
> rf_SC$model.


Random Forest.
7620 samples.
  20 predictor.
    6 classes: 'Bare areas', 'Built-up', 'Cropland', 'Grass/ open', 'Water', 'Woodlands'.
No pre-processing.
Resampling: Cross-Validated (fivefold).
Summary of sample sizes: 6096, 6095, 6097, 6095, 6097.
Resampling results across tuning parameters:

  mtry       Accuracy       Kappa.
  2          0.7484217      0.6968044.
  11         0.7468465      0.6948800.
  20         0.7408074      0.6876037.


Accuracy was used to select the optimal model using the largest value.
The final value used for the model was mtry = 2.
```

The results show that 7,620 training samples were used for training. The 20 predictors represent Sentinel-1 mean and median VV and VH features as well as texture indices. The six land cover classes represent the response (target) variable. The best model had an mtry value of 2 with an overall accuracy of 74.8% (Fig. 4.6).

Next, display the RF model training performance (Fig. 4.6).

```
plot(rf_SC$model) # Plot CV model.
```

Check the parameters of the best model.

Fig. 4.6 Repeated cross-validation (based on overall accuracy) profile for the RF model

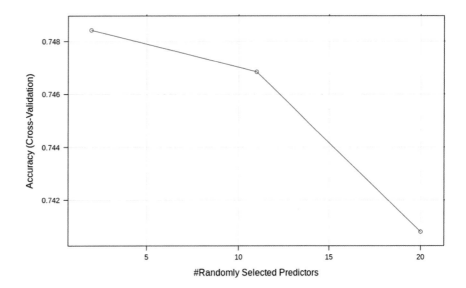

```
> # Check the parameters of the best model.
> rf_SC$model$finalModel.
```

```
Call:
 randomForest(x = x, y = y, mtry = param$mtry)
                    Type of random forest: classification
                          Number of trees: 500
No. of variables tried at each split: 2
                    OOB estimate of error rate: 24.7%
```

Confusion matrix:

	Bare areas	Built-up	Cropland	Grass/open	Water	Woodlands	class.error
Bare areas	747	222	68	135	10	38	0.38770492
Built-up	53	1047	37	136	0	162	0.27038328
Cropland	50	39	1181	228	0	17	0.22046205
Grass/open	84	83	273	833	0	48	0.36941711
Water	6	1	3	1	1016	1	0.01167315
Woodlands	6	117	3	61	0	914	0.16984559

The output shows a confusion matrix for the best model (after cross-validation). A total of 500 decision trees were used in the RF model. From the 20 predictors, only two predictors were selected at each split. The out-of-bag (OOB) estimate of error rate is 24.7%.

Next, display variable importance using ***varImp()*** function (Fig. 4.7).

```
# Compute Variable Importance.
SC_varImp<- varImp(rf_SC$model, compete=FALSE).
ggplot(SC_varImp, top=10).
```

Figure 4.7 shows the relative importance of the top 10 predictors. The top four predictors are mean VV from the post-rainy season (S1_mean_Asc_VV_Apr_Jun_20), and mean VH texture indices (mean_S1_VH_rainy_20), mean VV (S1_mean_Asc_VV_Jan_Mar_20), and mean VH (S1_mean_Asc_VH_Jan_Mar_20) from the rainy season. Note that all the predictors are from the rainy and post-rainy seasons.

Fig. 4.7 Random forest variable importance

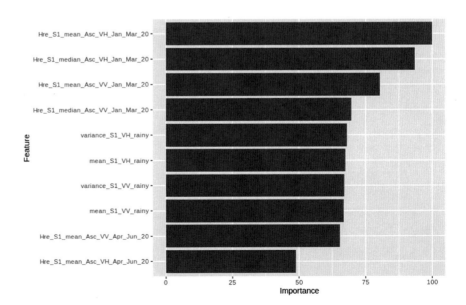

Next, perform validation of the random forest model. First, we use the RF model results for prediction and then build a confusion matrix.

```
> # Checkmodel validation.
> rf_SC.

superClass results.
*********** Validation **************
$validation.
Confusion Matrix and Statistics.
```

Reference

Prediction	Bare areas	Built-up	Cropland	Grass/ open	Water	Woodlands
Bare areas	**570**	35	43	122	11	20
Built-up	288	**779**	16	49	0	115
Cropland	78	5	**922**	240	0	8
Grass/ open	107	68	202	**654**	0	111
Water	8	0	0	0	**1001**	0
Woodlands	34	202	18	36	0	**782**

```
Overall Statistics
Accuracy              : 0.7216
95% CI                : (0.7106, 0.7325)
No Information Rate    : 0.1841
P-Value [Acc > NIR]   : < 2.2e-16
Kappa                 : 0.6657
Mcnemar's Test P-Value : NA
Statistics by Class:
```

	Class: Bare areas	Class: Built-up Class:	Cropland Class:	Grass/ open Class:	Water Class:	Woodlands
Sensitivity	0.52535	0.7153	0.7677	0.5940	0.9891	0.7548
Specificity	0.95753	0.9139	0.9378	0.9100	0.9985	0.9472
Pos Pred Value	0.71161	0.6247	0.7358	0.5727	0.9921	0.7295
Neg Pred Value	0.91001	0.9413	0.9471	0.9169	0.9980	0.9534
Prevalence	0.16631	0.1669	0.1841	0.1688	0.1551	0.1588
Detection Rate	0.08737	0.1194	0.1413	0.1002	0.1534	0.1199
Detection Prevalence	0.12278	0.1911	0.1921	0.1750	0.1547	0.1643
Balanced Accuracy	0.74144	0.8146	0.8528	0.7520	0.9938	0.8510

The confusion matrix shows that the overall classification accuracy is 72%, which is much better than lab 1 but lower than Sentinel-2 derived land cover classification in Chap. 3. However, serious classification problems are also observed. Analysis of individual accuracies reveals important insights. The producer's accuracy (sensitivity) is lower than the user's accuracy (Pos Pred Value) for the bare areas. This indicates high omission errors, and therefore, the RF model significantly underestimated the bare areas. However, the producer's accuracy (sensitivity) is higher than the user's accuracy (Pos Pred Value) for the built-up class indicating that the RF model overestimated built-up areas. In contrast, the cropland class has high

Fig. 4.8 Land cover classification based on multi-seasonal Sentinel-1 imagery, texture indices, and a random forest classifier

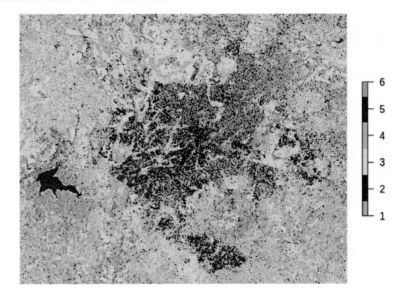

producer's and user's accuracies, and hence low omission and commission errors. For the grass/open areas class, the producer's accuracy is relatively higher than the user's accuracy indicating high commission errors. With respect to the woodland class, the producer's accuracy is higher than the user's accuracy. This indicates high commission errors and thus an overestimation of woodland class. The water class has the highest individual accuracies, which indicates that the RF model produced an optimal classification. While the RF model had difficulty in separating built-up and bare areas and cropland and grass/open areas classes, there is a slight improvement in accuracy. This suggest that additional texture indices improved land cover classification.

Display the land cover map. The land cover map also show poor accuracy, especially for the built-up class (Fig. 4.8).

```
> # Display the land cover map.
> plot(rf_SC$map, col=colors, legend=FALSE, axes=FALSE, box=FALSE).
> legend(1,1, legend=levels(ta_data$Class), fill=colors, title="Classes", horiz=TRUE, bty="n").
```

Next, we are going to apply a majority filter based on a 3 × 3 window filter in order to remove small pixels that cause a salt-and-pepper effect on the land cover map.

```
> # Filter the land cover map.
> # Create a 3×3 window filter.
> window<- matrix(1, 3, 3).

> # Perform majority filtering.
> RF_2020_MF<- focal(rf_SC$map, w=window, fun=modal)
```

Step 5: Save the final land cover map.

Finally, save the land cover map so that it can be displayed in QGIS or other GIS software.

```
# Save land cover map.
writeRaster(RF_2020_MF, filename="S1_MS_Texture_LC_Jan_Sep_2020b.img", type="raw",
            datatype='INT1U',index=1, na.rm=TRUE, progress="window", overwrite=TRUE).
```

4.3 Summary

In this chapter, we completed two land cover mapping labs using multi-seasonal Sentinel-1 imagery and texture indices (derived from Sentinel-1 imagery). In lab 1, we used only multi-seasonal Sentinel-1 imagery, while in lab 2, multi-seasonal Sentinel-1 and texture indices were used. The land cover map produced in the second lab had a relatively higher accuracy than the land cover map produced in lab 1 (73% versus 68%). The improvement in accuracy attributed to additional texture indices derived from Sentinel-1 imagery. While an improvement has been noted, significant problems were observed. Nonetheless, results show that the value of texture indices derived from Sentinel-1 imagery. This suggests that a fusion of Sentinel-1 and Sentinel-2 imagery as well as other variables derived from these satellite data may minimize spectral confusion, and therefore improve land cover classification accuracy.

4.4 Additional Exercises

i   Produce land cover maps using seasonal Sentinel-1 imagery, spectral and texture indices, and support vector machines (SVM).
ii Compare SVM and RF classification results.
iii Discuss the performance of the SVM and RF classifiers in terms of model performance and validation.

References

Bryan ML (1979) The effect of radar azimuth angle on cultural data. Photogramm Eng Remote Sens V 45(8):1097–1107

Corbane C, Faure JF, Baghdadi N, Villeneuve N, Petit M (2008) Rapid urban mapping using SAR/Optical imagery synergy. Sensors 8:7125–7143. https://doi.org/10.3390/s8117125

Dell'Acqua F, Gamba P (2003) Texture-based characterization of urban environments on satellite SAR images. IEEE Trans Geosci Remote Sens 41(1):153–159

Eckardt R, Urbazaev M, Salepci N, Pathe C, Schmullius, C, Woodhouse I, Stewart C (2019) Echoes in space: introduction to radar remote sensing. European Space Agency and EO College

ESA (2015) Sentinel-2 user handbook. Available at https://doi.org/10.13128/REA-22658. Accessed on 1 Aug 2020

Flores A, Herndon K, Thapa R, Cherrington E (eds) (2019) SAR handbook: comprehensive methodologies for forest monitoring and biomass estimation NASA 2019 https://doi.org/10.25966/1rgr-q551

Goldblatt R, Klaus Deininger K, Hanson G (2018) Utilizing publicly available satellite data for urban research: mapping built-up land cover and land use in Ho Chi Minh City. Vietnam. Development Engineering 3:83–99

Griffiths P, Hostert P, Gruebner O, van der Linden S (2010) Mapping megacity growth with multi-sensor data. Remote Sens Environ 114:426–439

Hardaway G, Gustafs GC, Lichy D (1982) Cardinal effect on seasat images of urban areas. Photogramm Eng Remote Sensing. 48(3):399–404

Jensen JR (2000) Remote sensing of the environment: an earth resource perspective. Pearson Education, pp 285–331

Kamusoko C, Kamusoko OW, Chikati E, Gamba J (2021) Mapping urban and peri-urban land cover in Zimbabwe: challenges and prospects. Geomatics 1:114–147

Lu D, Weng Q (2007) A survey of image classification methods and techniques for improving classification performance. Int J Remote Sens 28:823–870

Mather PM, Koch M (2011) Computer processing of remotely-sensed images: an introduction. Wiley-Blackwell, Chichester

Roberts SA, Robertson C (2016) Geographic information systems and science: a concise handbook of spatial data handling, representation, and computation. Oxford, pp 23–24

Sabins FF (1997) Remote sensing: principles and interpretation. W. H. Freeman and Company, pp 177–240

Schneider A (2012) Monitoring land cover change in urban and peri-urban areas using dense time stacks of Landsat satellite data and a data mining approach. Remote Sens Environ 124:689–704

Schug F, Okujeni A, Hauer J, Hostert P, Nielsen JØ, van der Linden S (2018) Mapping patterns of urban development in Ouagadougou, Burkina Faso, using machine learning regression modeling with bi-seasonal Landsat time series. Remote Sens Environ 210:217–228

Seto KC, Fragkias M, Güneralp B, Reilly MK (2011) A Meta-analysis of global urban land expansion. PLoS ONE 6(8):e23777. https://doi.org/10.1371/journal.pone.0023777

Tso B, Mather PM (2001) Classification methods for remotely sensed data. Taylor & Francis, pp 24–53

Zhu Z, Woodcock CE, Rogan J, Kellndorfer J (2012) Assessment of spectral, polarimetric, temporal, and spatial dimensions for urban and peri-urban land cover classification using Landsat and SAR data. Remote Sens Environ 117:72–82

Improving Urban Land Cover Mapping

5

Abstract

Multispectral optical and SAR imagery have potential to improve land cover mapping in urban areas. This is because satellite imagery such as Sentinel-1 and Sentinel-2 has high spatial and temporal resolutions. Therefore, multi-seasonal optical imagery can be used to discriminate built-up areas from cropland and bare areas, while SAR imagery can be used to detect built-up structures. In addition, spectral and texture indices derived from Sentinel-2 and Sentinel-1 imagery can be used to improve urban land cover mapping. In this chapter, Sentinel-1 and Sentinel-2 imagery as well as spectral and texture indices will be used for land cover mapping.

Keywords

SAR • Sentinel-1 • Sentinel-2 • Multi-seasonal • Spectral indices • Texture indices • Land cover mapping

5.1 Background

Mapping urban land cover is challenging due to the spectral confusion and mixed-pixel problems (Griffiths et al. 2010, Maktav and Erbek 2005). This is because built-up areas are composed of concrete, asphalt, metal, and shingles that are spectrally similar to bare areas and fallow croplands (Scheineder 2012; Kamusoko et al. 2013). Furthermore, medium to high spatial resolution satellite imagery pixels are usually mixtures of several land cover types (Schug et al. 2018). As a result, built-up areas are underestimated, especially in low density urban and peri-urban areas (Kamusoko et al. 2021). It is also challenging to map land cover in informal settlements characterized by built-up structures of different shapes, sizes, and orientation. Many techniques such as supervised and sub-pixel classifications (Lo and Choi 2004), object-based classifications (Guidon et al. 2004), spectral and texture indices (Moller-Jensen 1990; Gong and Howarth 1990; Xu 2007), and fusion of multiple satellite data (Peresi et al. 2016; Goldblatt et al. 2018a) have been used to improve land cover mapping. It is also noteworthy to point out that most urban land cover mapping studies have be done in megacities and large urban centers located in China, North America, and Europe (Griffiths et al. 2010; Zhu et al. 2019). However, the fastest growing small and medium urban areas with less than 1 million inhabitants (Bello-Schuneman et al. 2018; Güneralp et al. 2017)—which accounts for about 59% of the global urban population—are poorly quantified.

The availability of synthetic aperture radar (SAR) Sentinel-1 and optical Sentinel-2 data provides a great opportunity to address some of the land cover mapping challenges. This is because Sentinel-1 and Sentinel-2 data have high spatial and temporal resolutions. Although previous studies have shown the utility of Sentinel-2 data for land cover mapping (Schug et al. 2020; Forkuor et al. 2017; Peresi et al. 2016), few studies have combined Sentinel-1 and Sentinel-2 for land cover mapping in urban areas (Goldblatt et al. 2018b; Haas and Ban 2017; Sinha et al. 2020). Therefore, there is need to explore how the combination of Sentinel-1 and optical Sentinel-2 data improves land cover mapping in urban areas. This is important because previous studies have shown that multi-seasonal optical data is useful for identifying phonological changes (Yuan et al. 2005), while the SAR sensor can detect high backscatter of built-up structures (Zhu et al. 2012). In particular, it is easy to detect high density urban areas oriented to the SAR sensor (Goldblatt et al. 2018b). However, SAR sensor scattering behavior is affected by its system properties and surface or target parameters such as topographic effects (Li et al. 2020). The scattering effects tend to complicate backscatter analysis in urban areas (Molch 2009). In this chapter, we are going to map land cover using Sentinel-1 and Sentinel-2 imagery as well as spectral and texture indices.

5.2   Land Cover Mapping Labs

5.2.1   Lab 1. Mapping Land Cover Using Multi-Seasonal Sentinel-1 and Sentinel-2 Imagery

Objective

- Map land cover using seasonal Sentinel-1 and Sentinel-2 imagery.

Procedure

In this lab, we are going to use Sentinel-1 and Sentinel-2 imagery.

Step 1: Load the required packages

```
> # First, load the following libraries
> require(sp)
> require(rgdal)
> require(raster)
> require(caret)
> require(randomForest)
> require(ranger)
> require(RStoolbox)
> require(ggplot2)
> require(ranger)
```

Step 2: Load the raster data

Set up the working directory that contains Sentinel-1 and Sentinel-2 imagery.

```
> # Set up the working directory:
> setwd("/home/kamusoko/Documents/Projects/DL_ML_Apps/GEE_Classifications/Urban/Harare/SatelliteImagery/
Sentinel/2020/Sentinel-2/Sentinel-2_All_Bands/Chapter 5")
```

Import multi-seasonal Sentinel-1 and Sentinel-2 imagery.

```
> # Create a list of raster bands that will be used for classification.
> rlist=list.files(getwd(),pattern="tif$", full.names=TRUE)
```

Next, combine the raster layers.

```
> # Combine or stack the raster layers.
> rvars <- stack(rlist)
```

Check the attributes of the raster object.

```
# Check the attributes of rvars
> rvars
class       : RasterStack
dimensions  : 4806, 5950, 28595700, 39 (nrow, ncol, ncell, nlayers)
```

```
resolution  : 10, 10  (x, y)
extent      : 260350, 319850, 8001640, 8049700  (xmin, xmax, ymin, ymax)
crs         : +proj=utm +zone=36 +south +datum=WGS84 +units=m +no_defs +ellps=WGS84
names       : Hre_S1_me//Apr_Jun_20, Hre_S1_me//Jan_Mar_20, Hre_S1_me//Jul_Sep_20, Hre_S1_me//Apr_Jun_20,
Hre_S1_me//Jan_Mar_20, Hre_S1_me//Jul_Sep_20.
```

Step 3: Load the training points
Load the shapefile that contains the training data points using the ***readOGR()*** function.

```
> # Read training data shapefile.
> ta_data <- readOGR(getwd(), "Hre_Revised_Polygon_TA_2020")
OGR data source with driver: ESRI Shapefile
Source: "/home/kamusoko/Documents/Projects/DL_ML_Apps/GEE_Classifications/Urban/Harare/SatelliteImagery/
Sentinel/2020/Sentinel-2/Sentinel-2_All_Bands/Chapter5/Lab1", layer: "Hre_Revised_Polygon_TA_2020"
with 5711 features
It has 1 fields
```

Next, check the attributes of the "ta_data", object.

```
> # Check the land cover class distribution.
> summary(ta_data)
Object of class SpatialPolygonsDataFrame
Coordinates:
       min       max
x 260516.9   319789.7
y 8001673.2  8049677.8
proj4string :
[+proj=utm +zone=36 +south +datum=WGS84 +units=m +no_defs +ellps=WGS84 +towgs84=0,0,0]
Data attributes:
     Class
 Bare areas  :1091
 Built-up    :2113
 Cropland    :1008
 Grass/ open :1095
 Water       : 73
 Woodlands   : 331
```

Step 4: Train the random forest (RF) model
We are going to define the RF model parameters using the *superClass()* function from **RSToolbox** package.

```
> ## Fit random forest model (splitting training into 70% training data, 30% validation data)
> # Set a pre-defined value using set.seed()
> hre_seed <- 27
> set.seed(hre_seed)
> # Set-up timer to check how long it takes to run the model
> timeStart <- proc.time()
> # Set-up the model parameters and run it.
> rf_SC  <- superClass(rvars, trainData = ta_data, responseCol = "Class",
+              model = "rf", tuneLength = 3, trainPartition = 0.7)
> proc.time() - timeStart
   user    system   elapsed
1912.843  30.400   1944.145
```

It took approximately about 30 min to train the RF model with the 39 predictors (Sentinel-2 bands and Sentinel-1 features). Next, check the RF model performance.

```
> # Check RF model parameters
> rf_SC$model
```

```
Random Forest
7620 samples
  39 predictor
  6 classes: 'Bare areas', 'Built-up', 'Cropland', 'Grass/ open', 'Water', 'Woodlands'
No pre-processing
Resampling: Cross-Validated (5 fold)
Summary of sample sizes: 6096, 6095, 6097, 6095, 6097
Resampling results across tuning parameters:
  mtry   Accuracy     Kappa
   2     0.9212611    0.9050968
  20     0.9186363    0.9019550
  39     0.9135171    0.8957897
Accuracy was used to select the optimal model using the largest value.
The final value used for the model was mtry = 2.
```

The results show that 7620 training samples were used for training. The 39 predictors are the Sentinel-1 and Sentinel-2 features, while the six land cover classes represent the response (target) variable. The best model had an mtry value of 2 with an overall accuracy of 92%, which is relatively good (Fig. 5.1).

Next, display the RF model training performance (Fig. 5.1).

```
# Plot CV model
plot(rf_SC$model)
```

Next, check the parameters of the best model.

```
> # Check the parameters of the best model.
> rf_SC$model$finalModel
Call:
 randomForest(x = x, y = y, mtry = param$mtry)
               Type of random forest: classification
                     Number of trees: 500
No. of variables tried at each split: 2
        OOB estimate of  error rate: 7.6%
Confusion matrix:
```

	Bare areas	Built-up	Cropland	Grass/ open	Water	Woodlands	class.error
Bare areas	1028	117	35	37	0	3	0.1573770492
Built-up	33	1363	28	10	0	1	0.0501742160
Cropland	18	25	1390	81	0	1	0.0825082508
Grass/ open	25	24	105	1154	0	13	0.1264193793
Water	0	0	0	0	1027	1	0.0009727626
Woodlands	0	3	2	17	0	1079	0.0199818347

The output shows a confusion matrix for the best model (after cross-validation). A total of 500 decision trees were used in the RF model. From the 39 predictors, only two predictors were selected at each split. The out-of-bag (OOB) estimate of error rate is 7.6%, which is much better than the previous RF models in Chaps. 3 and 4.

Next, display variable importance using *varImp()* function (Fig. 5.2).

Fig. 5.1 Repeated
cross-validation (based on overall
accuracy) profile for the RF
model

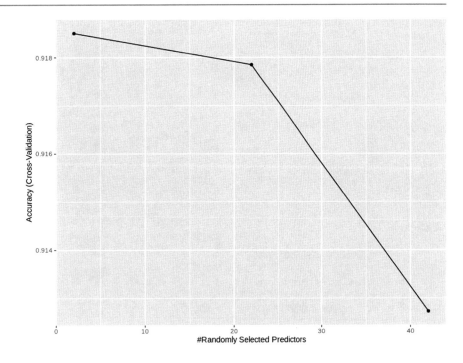

Fig. 5.2 Random forest
(RF) variable importance

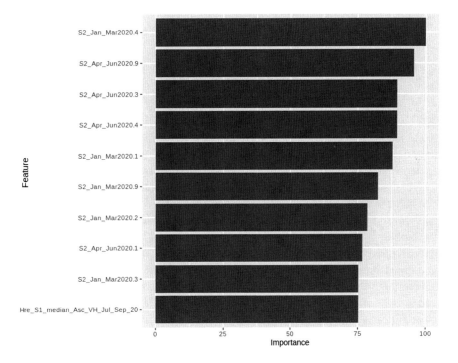

```
> # Compute Variable Importance
> SC_varImp <- varImp(rf_SC$model, compete = FALSE)
> ggplot(SC_varImp, top = 10)
```

Figure 5.2 shows the contribution of the top ten predictors. The top nine predictors are all from the rainy and post-rainy seasons. The results are consistent with spectral profiles shown in Fig. 3.1.2a and b, which indicate that rainy and post-rainy season imagery has potential to improve land cover mapping in the study area. The median VH band from the dry season is also within the top ten predictors. This suggests that land cover mapping in the study area should also include Sentinel-1 imagery.

Next, perform validation of the RF model. First, we use the model results for prediction and then build a confusion matrix.

```
> # Checkmodel validation
> rf_SC
*********** Validation **************
$validation
Confusion Matrix and Statistics
        Reference
Prediction          Bare areas      Built-up      Cropland      Grass/ open      Water      Woodlands
  Bare areas         717            46            46            46               0          0
  Built-up           221            1020          18            25               0          0
  Cropland           95             6             1072          239              0          0
  Grass/ open        52             11            65            780              0          11
  Water              0              3             0             0                1012       1
  Woodlands          0              3             0             11               0          1024
Overall Statistics
              Accuracy : 0.8622
                95% CI : (0.8536, 0.8705)
    No Information Rate : 0.1841
    P-Value [Acc > NIR] : < 2.2e-16
                 Kappa : 0.8344
Statistics by Class:
                      Class: Bare     Class:        Class:        Class:         Class:      Class:
                      areas           Built-up      Cropland      Grass/ open    Water       Woodlands
Sensitivity           0.6608          0.9366        0.8926        0.7084         1.0000      0.9884
Specificity           0.9746          0.9514        0.9361        0.9744         0.9993      0.9974
Pos Pred Value        0.8386          0.7944        0.7592        0.8487 ,       0.9961      0.9865
Neg Pred Value        0.9351          0.9868        0.9748        0.9427         1.0000      0.9978
Prevalence            0.1663          0.1669        0.1841        0.1688         0.1551      0.1588
Detection Rate        0.1099          0.1563        0.1643        0.1196         0.1551      0.1570
Detection Prevalence  0.1311          0.1968        0.2164        0.1409         0.1557      0.1591
Balanced Accuracy     0.8177          0.9440        0.9144        0.8414         0.9996      0.9929
```

The overall classification accuracy is 86.2%, which is better than the previous labs. The correctly classified metric (diagonal areas in the error matrix) significantly improved for all land cover classes except for the bare areas class. There are high omission errors for the bare areas class as most built-up (221), cropland (95), and grass/ open areas (52) pixels are excluded. The built-up areas class include only 46 misclassified bare areas pixels. However, there is a substantial increase in the commission errors for the built-up class since it includes 221 misclassified bare areas pixels. This indicates that spectral confusion is still a problem despite the fact that we used multi-seasonal Sentinel-1 and Sentinel-2 imagery. With respect to the cropland class, a relatively small amount of pixels are excluded. In particular, there is relatively low confusion between the built-up and cropland areas. This is very significant because spectral confusion between cropland and built-up areas in optical imagery is usually the main cause of poor classification accuracy. However, there is a significant amount of grass/open areas that has been incorrectly included in the cropland class. For the grass/open areas, there is also a significant amount of cropland pixels that have been misclassified. This suggests that the RF model had difficulty to separate cropland and grass/open areas. Generally, woodland and water have relatively high accuracies. The improvement in the woodlands class indicates the value of combining multi-seasonal Sentinel-1 and Sentinel-2 data.

The class accuracy metrics reveal very important information, which may be used to improve the model. The producer's accuracy (sensitivity) is significantly lower than the user's accuracy (Pos Pred Value) for the bare areas. This indicates high omission errors, and thus, the RF model significantly underestimated the bare areas. With respect to the built-up class, the producer's accuracy (sensitivity) is significantly higher than the user's accuracy (Pos Pred Value). This shows that the RF model overestimated built-up areas given the high commission errors. Similarly, the producer's accuracy (sensitivity) is higher than the user's accuracy (Pos Pred Value) for the cropland class, which indicates relatively high commission errors and thus an overestimation of the cropland areas. In contrast, the producer's accuracy is relatively lower than the user's accuracy for the grass/open areas class. This indicates high errors of omission and thus an underestimation of the grass/ open areas. Generally, the woodland and water classes have high individual accuracies, which indicate an improvement in classification. While multi-seasonal Sentinel-2 and Sentinel-1 imagery improved land cover mapping, classification errors are still present. The RF model still has difficulties to separate built-up and bare areas on one hand and cropland and grass/ open areas on the other hand. However, it is noteworthy that spectral confusion between built-up areas and cropland areas is

Fig. 5.3 Land cover classification based on multi-seasonal Sentinel-1 and Sentinel-2 imagery

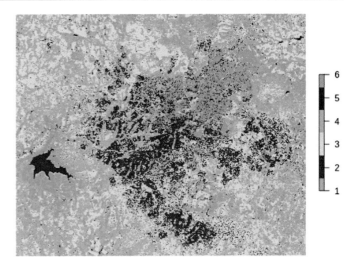

significantly minimized. This suggests that the combination of multi-seasonal Sentinel-1 and Sentinel-2 imagery had positive impact in the classification.

Display the land cover map (Fig. 5.3).

```
> # Display the land cover map
> colors <- c("grey","red", "yellow", "light green", "blue", "green")
> plot(rf_SC$map, col = colors, legend = TRUE, axes = FALSE, box = FALSE)
> legend(1,1, legend = levels(ta_data$Class), fill = colors , title = "Classes", horiz
    = TRUE,  bty = "n")
```

We are going to apply a majority filter based on a 3 × 3 window filter in order to remove small pixels that cause a salt and pepper effect on the land cover map (Fig. 5.3).

```
> # Filter the land cover map.
> # Create a 3x3 window filter
> window <- matrix(1, 3, 3)

> # Perform majority filtering
> RF_2020_MF <- focal(rf_SC$map, w = window, fun = modal)
```

Step 5: Save the final land cover map
Finally, save the land cover map.

```
# Save land cover map
writeRaster(RF_2020_MF, filename="S1_S2_MS_LC_Jan_Sep_2020a.img",
      type="raw", datatype='INT1U',index=1,na.rm=TRUE,progress="window",
      overwrite=TRUE)
```

5.2.2   Lab 2. Mapping Land Cover Using Multi-Seasonal Sentinel-1 and Sentinel-2 Imagery and Other Derived Data

Objective

- Map land cover using multi-seasonal Sentinel-1 and Sentinel-2 imagery and other variables.

Procedure

In this lab, we are going to use multi-seasonal Sentinel-1 and Sentinel-2 imagery and other variables (spectral and texture indices).

Step 1: Load the required packages

```
> # First, load the following libraries
> require(sp)
> require(rgdal)
> require(raster)
> require(caret)
> require(randomForest)
> require(ranger)
> require(RStoolbox)
> require(ggplot2)
> require(ranger)
```

Step 2: Load the raster data
Set up the working directory where all data sets are located.

```
> # Set up the working directory:
> setwd("/home/kamusoko/Documents/Projects/DL_ML_Apps/GEE_Classifications/Urban/Harare/SatelliteImagery/
Sentinel/2020/Sentinel-2/Sentinel-2_All_Bands/Chapter 5")
```

Import multi-seasonal Sentinel-1 imagery, Sentinel-2 imagery, and spectral and texture indices.

```
> # Create a list of raster bands that will be used for classification.
> rlist=list.files(getwd(),pattern="tif$", full.names=TRUE)
```

Next, combine the raster layers.
Combine or stack the raster layers.

```
> # Combine or stack the raster layers.
> rvars <- stack(rlist)
```

Check the attributes of raster object.

```
> # Check the attributes of Sentinel-1 imagery.
> rvars
class       : RasterStack
dimensions  : 4806, 5950, 28595700, 94  (nrow, ncol, ncell, nlayers)
resolution  : 10, 10  (x, y)
```

```
extent      : 260350, 319850, 8001640, 8049700 (xmin, xmax, ymin, ymax)
crs         : +proj=utm +zone=36 +south +ellps=WGS84 +towgs84=0,0,0,0,0,0,0 +units=m +no_defs
names       : entropy_S1_VH_rainy, entropy_S1_VV_rainy, entropy_S1_VV_VH_rainy, homogene-
ity_S1_VH_rainy, homogeneity_S1_VV_rainy, homogeneity_S1_VV_VH_rainy, Hre_mean_//r_Jun_2020, Hre_mean_//
n_Mar_2020, Hre_mean_//Jul_Sep_20, Hre_S1_me//Apr_Jun_20.
```

Step 3: Load the training points

Load the shapefile that contains the training data points using the ***readOGR()*** function.

```
> # Read training data shapefile.
> ta_data <- readOGR(getwd(), "Hre_Revised_Polygon_TA_2020")
OGR data source with driver: ESRI Shapefile
Source: "/home/kamusoko/Documents/Projects/DL_ML_Apps/GEE_Classifications/Urban/Harare/SatelliteImagery/
Sentinel/2020/Sentinel-2/Sentinel-2_All_Bands/Chapter5/Lab1", layer: "Hre_Revised_Polygon_TA_2020"
with 5711 features
It has 1 fields
```

Next, check the attributes of the "ta_data", object.

```
> # Check the land cover class distribution.
> summary(ta_data)
Object of class SpatialPolygonsDataFrame
Coordinates:
      min        max
x    260516.9   319789.7
y    8001673.2  8049677.8
Is projected: TRUE
proj4string :
[+proj=utm +zone=36 +south +datum=WGS84 +units=m +no_defs +ellps=WGS84 +towgs84=0,0,0]
Data attributes:
      Class
 Bare areas   :1091
 Built-up     :2113
 Cropland     :1008
 Grass/ open  :1095
 Water        : 73
 Woodlands    : 331
```

Step 4: Train the random forest (RF) model

We are going to define the RF model parameters.

```
> ## Fit random forest model (splitting training into 70% training data, 30% validation data)

> # Set a pre-defined value using set.seed()
> hre_seed <- 27
> set.seed(hre_seed)
> # Set-up timer to check how long it takes to run the model
> timeStart <- proc.time()
> # Set-up the model parameters and run it.
```

```
> rf_SC  <- superClass(rvars, trainData = ta_data, responseCol = "Class",
+                 model = "rf", tuneLength = 3, trainPartition = 0.7)
> proc.time() - timeStart
  user  system elapsed
2590.208  65.704 2659.188
```

It took approximately about 45 min to train the RF model with 108 predictors (Sentinel-2 bands, Sentinel-1 features, and other derived variables). Next, check the RF model performance.

```
> # Check RF model parameters
> rf_SC$model
Random Forest
7620 samples
 108 predictor
  6 classes: 'Bare areas', 'Built-up', 'Cropland', 'Grass/ open', 'Water', 'Woodlands'
No pre-processing
Resampling: Cross-Validated (5 fold)
Summary of sample sizes: 6096, 6095, 6097, 6095, 6097
Resampling results across tuning parameters:
  mtry  Accuracy   Kappa
   2    0.9297891  0.9153853
   55   0.9300512  0.9157178
   108  0.9249343  0.9095502
Accuracy was used to select the optimal model using the largest value.
The final value used for the model was mtry = 55.
```

The results show that 7,620 training samples were used for training. The 108 predictors are the Sentinel-1 and Sentinel-2 bands and spectral and texture indices. The six land cover classes represent the response (target) variable. The best model had an mtry value of 55 with an overall accuracy of 93%, which is relatively good (Fig. 5.4).

Next, display the RF model training performance (Fig. 5.4).

```
# Plot CV model
plot(rf_SC$model)
```

Next, check the parameters of the best model.

```
> # Check the parameters of the best model.
> rf_SC$model$finalModel
Call:
 randomForest(x = x, y = y, mtry = param$mtry)
          Type of random forest: classification
          Number of trees: 500
          No. of variables tried at each split: 55
        OOB estimate of  error rate: 6.81%
Confusion matrix:
```

	Bare areas	Built-up	Cropland	Grass/ open	Water	Woodlands	class. error
Bare areas	**1051**	98	33	33 0.138524590	1	4	
Built-up	31	**1376**	17	10 0.041114983	0	1	
Cropland	20	16	**1390**	88 0.082508251	0	1	
Grass/ open	34	19	79	**1177** 0.109008327	0	12	
Water	0	0	0	1 0.001945525	**1026**	1	
Woodlands	1	1	3	15 0.018165304	0	**1081**	

Fig. 5.4 Repeated
cross-validation (based on overall
accuracy) profile for the RF
model

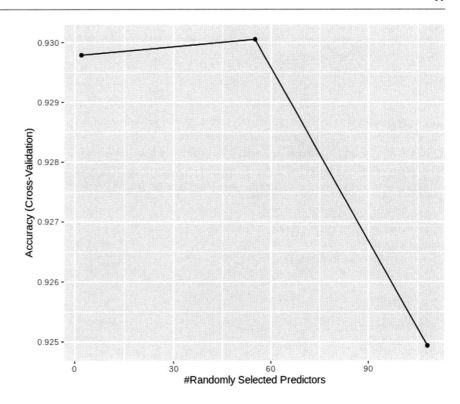

The output shows a confusion matrix for the best model (after cross-validation). A total of 500 decision trees were used in the RF model. From the 108 predictors, only 55 predictors were selected at each split. The out-of-bag (OOB) estimate of error rate is 6.8%, which is much better than the previous RF models.

Next, display variable importance using ***varImp()***function (Fig. 5.5).

Fig. 5.5 Random forest variable
importance

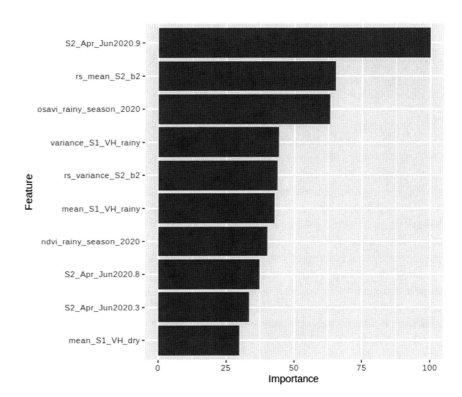

```
> # Compute Variable Importance
> SC_varImp <- varImp(rf_SC$model, compete = FALSE)
> ggplot(SC_varImp, top = 10)
```

Figure 5.5 shows the relative importance of the top ten predictors. The top three predictors are all from the rainy seasons. The first predictor is the vegetation red edge band 1 (S2_Jan_March.4), the second is mean VH Sentinel-1 band (mean_SI_VH_rainy), and third is mean VH Sentinel-1 band (variance_SI_VH_rainy). Furthermore, rainy and post-rainy seasons OSAVI, NDVI, and VV variance as well as mean VH band from the dry season have significant impact in the RF model. This shows the importance of using Sentinel-1 and Sentinel-2 derived variables for land cover mapping in the study area.

Next, perform RF model validation. First, we use the model results for prediction and then build a confusion matrix.

```
superClass results
************ Validation **************
$validation
Confusion Matrix and Statistics
          Reference
```

Prediction	Bare areas	Built-up	Cropland	Grass/ open	Water	Woodlands
Bare areas	**757**	46	55	28	0	0
Built-up	181	**1026**	10	16	0	0
Cropland	75	3	**1050**	407	0	0
Grass/ open	65	5	84	**638**	0	15
Water	7	4	0	0	**1012**	5
Woodlands	0	5	2	12	0	**1016**

```
Overall Statistics
              Accuracy : 0.8429
                95% CI : (0.8338, 0.8516)
    No Information Rate : 0.1841
    P-Value [Acc > NIR] : < 2.2e-16
                 Kappa : 0.8112
 Mcnemar's Test P-Value : NA
Statistics by Class:
```

	Class: Bare areas	Class: Built-up	Class: Cropland	Class: Grass/ open	Class: Water	Class: Woodlands
Sensitivity	0.6977	0.9421	0.8743	0.57947	1.0000	0.9807
Specificity	0.9763	0.9619	0.9089	0.96884	0.9971	0.9965
Pos Pred Value	0.8544	0.8321	0.6840	0.79058	0.9844	0.9816
Neg Pred Value	0.9418	0.9881	0.9697	0.91901	1.0000	0.9964
Prevalence	0.1663	0.1669	0.1841	0.16876	0.1551	0.1588
Detection Rate	0.1160	0.1573	0.1609	0.09779	0.1551	0.1557
Detection Prevalence	0.1358	0.1890	0.2353	0.12370	0.1576	0.1586
Balanced Accuracy	0.8370	0.9520	0.8916	0.77415	0.9985	0.9886

The confusion matrix shows that the overall classification accuracy is 84%, which is slightly lower than the previous results in lab 1 (86%). The correctly classified metric significantly improved for the bare areas and built-up areas, while it decreased for cropland, grass/open areas, and woodlands. However, the omission errors for the bare areas class are still high as most built-up (181), cropland (75), and grass/ open areas (65) pixels are excluded. For the built-up areas class, a slight decrease in commission errors is observed compared to the previous results in lab 1. The built-up class include 181 misclassified bare areas pixels. This indicates a slight improvement, which suggests that multi-seasonal Sentinel-1 and Sentinel-2 and spectral and texture indices had impact on bare areas and built-up areas. With respect to the cropland class, a slight increase in omission errors is observed as most bare areas (55) and grass/open areas (84) pixels are excluded. However, spectral confusion is minimized between the built-up and cropland classes. This is suggested that multi-seasonal Sentinel-1 and Sentinel-2, spectral indices and texture had impact on built-up and cropland areas. Nonetheless, there is a significant amount of grass/open areas that has been incorrectly included in cropland class. For the grass/open areas, there is also an increased amount of cropland pixels that have been misclassified. As noted in the previous lab, the RF model had difficulty to separate cropland and grass/open areas. Generally, woodland and water have relatively high accuracies.

Fig. 5.6 Land cover
classification based on
multi-seasonal Sentinel-2 imagery
and other variables

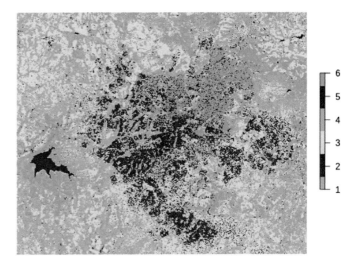

The class accuracy metrics reveal very important insights. As was observed in the previous lab, the producer's accuracy (sensitivity) is significantly lower than the user's accuracy (Pos Pred Value) for the bare areas. This indicates high omission errors, and thus, the RF classifier significantly underestimated bare areas. Although the built-up areas class have high individual accuracies, the producer's accuracy (sensitivity) is relatively higher than the user's accuracy (Pos Pred Value). This indicates an overestimation of built-up areas. With respect to the cropland class, the producer's accuracy (sensitivity) is significantly higher than the user's accuracy (Pos Pred Value) indicating high commission errors and thus an overestimation of this class. For the grass/open areas class, the producer's accuracy is substantially lower than the user's accuracy indicating high omission errors and thus an underestimation of this class. This is attributed to spectral confusion between grass/open areas and cropland areas. Generally, the woodland and water classes have high individual accuracies. It is noteworthy that class accuracies for built-up areas increased substantially. As a result, spectral confusion between built-up areas and cropland and bare areas is significantly minimized. This suggests that spectral and texture indices derived from multi-seasonal Sentinel-1 and Sentinel-2 data improved the classification of built-up areas. However, high omission errors for the grass/open areas class significantly reduced overall classification accuracy. Interestingly, the use of additional spectral index based on the vegetation red edge failed to minimize spectral confusion between cropland and grass/open areas as was originally assumed.

Display the land cover map (Fig. 5.6).

```
> # Display the land cover map
> colors <- c("grey","red", "yellow", "light green", "blue", "green")
> plot(rf_SC$map, col = colors, legend = TRUE, axes = FALSE, box = FALSE)
> legend(1,1, legend = levels(ta_data$Class), fill = colors , title = "Classes", horiz
    = TRUE,  bty = "n")
```

We are going to apply a majority filter based on a 3 × 3 window filter in order to remove small pixels that cause a salt and pepper effect on the land cover map.

```
> # Filter the land cover map.
> # Create a 3x3 window filter
> window <- matrix(1, 3, 3)


> # Perform majority filtering
> RF_2020_MF <- focal(rf_SC$map, w = window, fun = modal)
```

Step 5: Save the final land cover map
Finally, save the land cover map.

```
# Save land cover map
writeRaster(RF_2020_MF, filename="S1_S2_SI_Texture_LC_Jan_Sep_2020b.img",
        type="raw", datatype='INT1U',index=1,na.rm=TRUE,progress="window",
        overwrite=TRUE)
```

5.3 Summary

In this chapter, we completed two land cover mapping labs using multi-seasonal Sentinel-1 and Sentinel-2 data and spectral and texture indices. In lab 1, we used only multi-seasonal Sentinel-1 and Sentinel-2 data, while in the lab 2, we used multi-seasonal Sentinel-1 and Sentinel-2 data as well as additional spectral and texture indices. The land cover map produced in the first lab had a slightly higher accuracy than the land cover map produced in lab 2 (86% versus 84%). While the overall classification accuracy was lower for lab 2, spectral confusion between built-up areas and cropland and bare areas was significantly minimized. This suggests that additional spectral and textural indices derived from Sentinel-1 and Sentinel-2 data improved the accuracy of built-up areas. Although spectral confusion is still a problem, results from the two labs show that a combination of optical and SAR data can significantly improve land cover mapping.

5.4 Additional Exercises

i.   Classify seasonal Sentinel-1 and Sentinel-2 imagery using support vector machines (SVM).
ii. Classify seasonal Sentinel-1 and Sentinel-2 imagery and spectral and texture indices using SVM.
iii. Compare the results from the RF and SVM classifiers.
iv. Discuss the performance of the SVM and RF classifiers in terms of model performance and validation.

References

Bello-Schuneman J (2018) Defining the future of Africa's brave new world. Afr Fact 45:12–18

Forkuor G, Dimobe K, Serme I, Tondoh JE (2017) Landsat-8 vs. Sentinel-2: examining the added value of sentinel-2's red-edge bands to land-use and land-cover mapping in Burkina Faso. GIScience Remote Sens. 1–24

Goldblatt R, Stuhlmacher MF, Tellman B, Clinton N, Hanson G, Georgescu M, Wang C, Serrano-Candela F, Khandelwal AK, Cheng WH, Balling RC Jr (2018a) Using Landsat and nighttime lights for supervised pixel-based image classification of urban land cover. Remote Sens Environ 205:253–275

Goldblatt R, Klaus Deininger K, Hanson G (2018b) Utilizing publicly available satellite data for urban research: mapping built-up land cover and land use in Ho Chi Minh City. Vietnam. Development Engineering. 3:83–99

Gong P, Howarth PJ (1990) The use of structural information for improving land-cover classification accuracies at the rural-urban fringe. Photogramm Eng Remote Sens 56(1):67–73

Griffiths P, Hostert P, Gruebner O, van der Linden S (2010) Mapping megacity growth with multi-sensor data. Remote Sens Environ 114:426–439

Güneralp B, Lwasa S, Masundire H, Parnell S, Seto KC (2017) Urbanization in Africa: Challenges and opportunities for conservation. *Environ. Res. Lett 13*(015002), doi:https://doi.org/10.1088/1748-9326/aa94fe.

Guindon B, Zhang Y, Dillabaugh C (2004) Landsat urban mapping based on a combined spectral-spatial methodology. Remote Sens Environ 92:218–232

Haas J, Ban Y (2017) Sentinel-1A SAR and sentinel-2A MSI data fusion for urban ecosystem service mapping. Remote Sens Appl: Soc Environ 8:41–53

Kamusoko C, Gamba J, Murakami H (2013) Monitoring urban spatial growth in Harare Metropolitan province Zimbabwe. Adv Remote Sens 2:322–331

Kamusoko C, Kamusoko OW, Chikati E, Gamba J (2021) Mapping urban and peri-urban land cover in Zimbabwe: Challenges and prospects. Geomatics (under review)

Li X, Zhou Y, Gong P, Seto KC, Clinton N (2020) Developing a method to estimate building height from Sentinel-1 data. Remote Sens Environ 240:1–8

Lo CP, Choi J (2004) A hybrid approach to urban land use/cover mapping using Landsat 7 enhanced thematic mapper plus (ETM+) images. Int J Remote Sens 25:2687–2700

Maktav D, Erbek FS (2005) Analysis of urban growth using multi-temporal satellite data in Istanbul Turkey. Int J Remote Sens 26(4):797–810

Molch K (2009) Radar Earth Observation Imagery for Urban Area Characterisation. https://doi.org/10.2788/8453

Moller-Jensen L (1990) Knowledge-based classification of classification of an urban area using texture and context information Landsat-TM imagery. Photogramm Eng Remote Sens 56(6):899–904

Pesaresi M, Corbane C, Julea A, Florczyk AJ, Syrris V, Soille P (2016) Assessment of the added-value of sentinel-2 for detecting built-up areas. Rem. Sens. 8:299. https://doi.org/10.3390/rs8040299

Schneider A (2012) Monitoring land cover change in urban and peri-urban areas using dense time stacks of landsat satellite data and a data mining approach. Remote Sens Environ 124:689–704

Schug F, Okujeni A, Hauer J, Hostert P, Nielsen JØ, van der Linden S (2018) Mapping patterns of urban development in Ouagadougou, Burkina Faso, using machine learning regression modeling with bi-seasonal Landsat time series. Remote Sens Environ 210:217–228

Schug F, Frantz D, Okujeni A, van der Linden S, Hostert P (2020) Mapping urban-rural gradients of settlements and vegetation at national scale using Sentinel-2 spectral-temporal metrics and regression-based unmixing with synthetic training data. Remote Sens Environ 246:111810

Sinha S, Santra A, Mitra SS (2020) Semi-automated impervious feature extraction using built-up indices developed from space-borne optical and SAR remotely sensed sensors. Adv Space Res xxx(2020):xxx–xxx

Xu H (2007) Extraction of urban built-up land features from Landsat imagery using a thematic-oriented index combination technique. Photogramm Eng Remote Sens 73:1381–1391

Yuan F, Sawaya KE, Loeffelholz BC, Bauer ME (2005) Land cover classification and change analysis of the Twin Cities (Minnesota) Metropolitan area by multitemporal Landsat remote sensing. Remote Sens Environ 98:317–328

Zhu Z, Woodcock CE, Rogan J, Kellndorfer J (2012) Assessment of spectral, polarimetric, temporal, and spatial dimensions for urban and peri-urban land cover classification using Landsat and SAR data. Remote Sens Environ 117:72–82

Zhu Z, Zhou Y, Seto KC, Stokes EC, Deng C, Pickett TA, Taubenböck H (2019) Understanding an urbanizing planet: Strategic directions for remote sensing. Remote Sens Environ 228:164–182

Land Cover Classification Accuracy Assessment

6

Abstract

Land cover mapping literature review is replete with accuracy assessment examples. However, most land cover maps lack rigorous accuracy assessment. This limits the scientific and real-world use of land cover maps. In order to overcome these limitations, recommended good practices are available in order to implement rigorous accuracy assessment. The good practice recommendations encourage estimation of area based on the reference classification and analysis of information contained in the confusion or error matrix. In this chapter, we are going to use the good practice recommendations based on sampling design, response design, and analysis.

Keywords

Land cover maps • Rigorous accuracy assessment • Good practices • Sampling design • Response design • Analysis

6.1 Background

During the past decades, the production of land cover maps has advanced rapidly given developments in Earth Observation (EO) sensor technology and machine learning techniques. Despite the rapid technological developments, land cover maps are still imperfect and contain errors (Foody 2010; Stehman and Foody 2019). As a result, accuracy assessment is required to communicate land cover map quality (Olofsson et al. 2014; Stehman and Foody 2019).

Generally, accuracy assessment is considered a major component of a land cover mapping project. Accuracy assessment entails that a land cover map is compared with higher quality reference data, which has been collected through a sample-based approach (Stehman 2009; Foody 2010). Based on this comparison, individual land cover class accuracy and overall accuracy metrics are produced. Although land cover mapping literature review is replete with accuracy assessment examples, most of the land cover maps produced have not been subjected to a rigorous evaluation (Olofsson et al. 2013; Stehman and Foody 2019; Morales-Barquero et al. 2019). Consequently, this limits the scientific and real-world utility of most land cover maps. In order to overcome these limitations, some researchers (Congalton and Green 1999; Foody 2002, 2010, 2020; Pontius and Millones 2011; Olofsson et al. 2013, 2014; Stehman and Foody 2019) designed guidelines to improve land cover map accuracy assessment. For example, Olofsson et al. (2014) and Stehman and Foody (2019) recommended the adoption of accuracy assessment good practices to improve the credibility of land cover maps. The good practice recommendations encourage estimation of area based on the reference classification and analysis of information contained in the confusion or error matrix (Olofsson et al., 2013, 2014).

In this workbook, we are going to follow the good practices based on the three major components of the good practices: (i) sampling design, (ii) response design, and (iii) analysis. Following is a brief description of the three major components of the good practices.

6.1.1 Sampling Design

Sampling design is the basis of accuracy assessment because it is expensive to collect reference data for the whole study area (Congalton and Green 1999; Foody 2010; Wulder et al. 2006). A sample refers to a subset of the study area or region of interest, while a sampling design is a protocol for selecting locations at which reference data will be collected (Stehman 2009). The critical points for selecting sampling design are that (i) the land cover classification scheme for the sampling data is the same as the one used in the classification; (ii) there is equal probability for all area to be sampled; and (iii) there is sufficient sampling size to achieve a proper statistical accuracy for each land cover class.

In general, the sample size for each land cover class is selected to ensure that the sample size is large enough to produce sufficiently precise estimates of the area of the class (Food and Agriculture Organization 2016). Therefore, a probability sampling design that includes randomization is a prerequisite since it quantifies the likelihood of inclusion. The likelihood of inclusion (inclusion probability) is the probability that a specific pixel is selected in the sample, which depends on the sampling design (Stehman 2009). Probability sampling designs include simple random, stratified random, and systematic sampling. In this workbook, we are going to use a stratified sampling approach, which has been recommended by Olofsson et al. (2014). Following are descriptions of critical sampling design steps.

6.1.2 (a) Stratified Random Sampling

A stratified random sampling procedure assigns specific number of sample points to each land cover class in proportion to the size or significance of the land cover class. The basic steps for stratified sampling design are to (i) divide all pixels among the land cover classes (which are the strata); (ii) take a random sample within each stratum and calculate accuracy estimates; (iii) combine the results from all the strata and then estimate the total number of correctly classified pixels within each stratum.

6.1.3 (b) Determine Sample Size

There are a variety of methods (and recommendations) to determine sample size. This is because the sample size depends on accuracy and area information that is not known prior to accuracy assessment. Generally, the analyst decides the best sample size. However, the sample size should be representative of the population, large enough to get reliable estimates, and cost-effective.

In this workbook, we are going to use the formula by Cochran (1977) in order to estimate the sample size. The formula takes as input the map areas for the classes to be assessed, a target standard error for overall accuracy, and expected user accuracies. The overall sample size will be distributed between the land cover classes taking into account proportional allocation of the land cover classes. That is, the overall sample size is allocated to the land cover class proportional to the area of the land cover class. In this regard, important land cover classes such as built-up area—which is the focus of the project—will have more proportion that small land cover classes such as water. However, small land cover classes should also have sufficient number of samples (Congalton and Green 2008).

6.1.4 Response Design

The response design comprises the framework that defines the agreement between the land cover map and the reference data (Olofsson et al. 2014). Here, it is important to establish reference data sources, which are going to be used to compare with the land cover map (Stehman and Foody 2019). While it is generally assumed that the reference data is more accurate than the land cover being assessed, it also important to note that the reference data might also contain errors (Foody 2002). Taking into account that reference data is usually collected from different sources (field surveys, high resolution imagery, or crowdsourcing) and that errors might be present, a detailed description of the response design is required in order to meet the transparency and reproducibility criteria (Olofsson et al. 2014; Stehman and Foody 2019). It is also important to provide documentation on the protocols used to define the land cover classes and ensure consistency among interpreters or teams of interpreters. In this workbook, we will focus on the following components of the response design: the spatial unit, the reference data sources, the labeling protocol for the reference classification, and the definition of agreement.

6.1.5 (a) Spatial Assessment Unit

The spatial assessment unit refers to the geographical entity on which comparison of the land cover map and the reference data is based (Foody 2002; Stehman and Foody 2019). This could be a pixel, group of pixels (e.g., block), or polygon (segments). Note that the spatial assessment unit must be selected with care because it has implications on the sampling design and analysis (Stehman and Foody 2019). In this workbook, the pixel will be used as the spatial assessment unit since the land cover map was produced using pixel-based machine learning approaches.

6.1.6 (b) Sources of Reference Data

To date, a variety of reference data sources that include field surveys, aerial photos, very high resolution satellite imagery, and crowdsourced data (Stehman et al. 2018) exist. A prerequisite is that reference data should be of higher quality compared to the land cover map. For example, if satellite imagery is to be used as a source of reference, then it should have higher spatial or radiometric resolution (Food and Agriculture Organization 2016). Alternatively, procedures to produce the reference data should be more accurate in cases when the same satellite imagery is used as reference data. This is usually the case in developing countries where very high resolution imagery may not be available for assessing the accuracy of land cover maps (Olofsson et al. 2014). In this workbook, we are going to use Google Satellite in QGIS as a source of the reference data.

6.1.7 (c) Reference Labeling Protocol

The labeling protocol defines the conversion of the reference data into a land cover class label, which will be used for comparison against the land cover map to be evaluated (Stehman and Foody 2019; Wulder et al. 2006). Therefore, the minimum mapping unit (MMU) and rules to allocate a land cover class label require careful consideration. This is because a large MMU may not adequately represent highly fragmented land cover mosaics (Stehman and Foody 2019). Furthermore, the same land cover definitions should be used for both land cover map under evaluation and the reference data (Congalton and Green 1999). Stehman and Foody (2019) also caution that a consistent approach for labeling reference classes should be established if reference data contains a mixture of land cover classes.

6.1.8 (d) Defining Agreement

Accuracy assessment requires the establishment of rules that define agreement between the land cover map and reference classifications. In general, if the both labels from the land cover map and reference classification agree, then the map is correct. However, this becomes complicated when a spatial unit covers more than one class. In this case, it is recommended to focus on the dominant land cover class in both the land cover map and reference data. For example, agreement can be defined as a match between the dominant (majority) land cover classes based on a 3 × 3 pixel neighborhood. If no dominant land cover class exists or if an accuracy assessment point lies between two land cover boundaries, then the pixel is marked as no confidence and therefore excluded from analysis.

6.1.9 Analysis

The analysis protocol focuses on how to tabulate the information contained in the comparison of land cover map and reference data into accuracy and area estimates as well as the quantification of uncertainty (Congalton and Green 2008; Olofsson et al. 2014; Stehman and Foody 2019). The confusion or error matrix is the basis of quantitative accuracy assessment metrics. This is because the confusion or error matrix provides a site-specific assessment between the land cover map and reference data. This section will focus briefly on the confusion matrix, estimating accuracy, and estimating area.

6.1.10 (a) The Confusion (Error) Matrix

The confusion matrix is a cross-tabulation of the class labels allocated by the land cover map and reference data (Congalton and Green 2008). Generally, the classified (predicted) land cover classes are represented in rows and the reference classes in columns. The main diagonal of the error matrix highlights correct classifications, while the off-diagonal elements show omission and commission errors. Olofsson et al. (2013) recommend that the confusion matrix should be reported in terms of proportions rather than counts. Furthermore, accuracy measures should be reported with their respective confidence intervals (Olofsson et al. 2013).

6.1.11 (b) Estimating Accuracy

The accuracy measures are derived from the error matrix and include overall accuracy, user's accuracy, producer's accuracy, area proportions, and their confidence intervals. The overall accuracy is the proportion of area classified correctly (Congalton and Green 2008; Olofsson et al. 2013). The user's accuracy is the proportion of the area mapped as a particular land cover class that is actually that land cover class on the ground (Congalton and Green 2008; Olofsson et al. 2013). The producer's accuracy is the proportion of area that is reference land cover class and is also land cover class in the map (Congalton and Green 2008; Olofsson et al. 2013). For all three accuracy measures, the confidence intervals need to be derived as well (Olofsson et al. 2014). However, the overall accuracy should be interpreted with care because it is not always representative of the accuracy of the individual classes or strata (FAO 2016). Although the kappa coefficient is also often reported as a measure of map accuracy, it is not recommended because it does not offer useful information (Pontius Jr and Millones 2011).

6.1.12 (c) Estimating Area

The confusion matrix will be used to estimate the area of land cover classes and their standard errors. The reference data will be used to adjust the area estimate as obtained from the map (Olofsson et al. 2014). The estimated area for each land cover class and the standard error of the estimated area is given by the equation in Olofsson et al. (2014).

6.2 Performing Accuracy Assessment

6.2.1 Lab 1. Sample Design

Objective

- To determine the sample size and allocation

  Prerequisite—Install the LecoS plugin.

Procedure

In this lab, we are going to determine the sample size and allocation.

1. Display your land cover map in QGIS by clicking *Layers > Add Raster Layer*.
2. Right-click the land cover map in the layer panel, and click *Properties > Properties > Symbology* (Fig. 6.1). Set *Render Type* to *Paletted/Unique values* and click *Classify*. Assign each land cover class an appropriate *Color* and *Label* (name).
3. Next, click the LecoS plugin, which we are going to use in order to calculate the area of each land cover class. LecoS refers to Landscape ecology statistics (Jung 2016).
4. To calculate the area of land cover class, go to **Raster > Landscape Ecology > Landscape statistics**. The **Landcover Analysis** panel appears (Fig. 6.2).

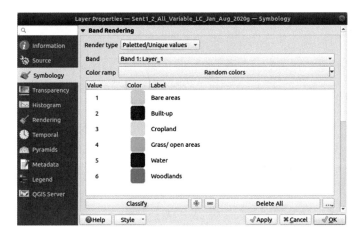

Fig. 6.1 Labeling land cover in QGIS

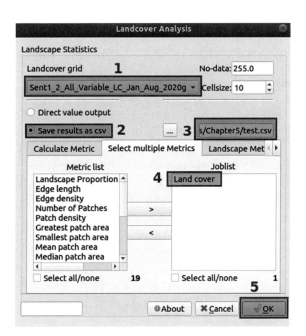

Fig. 6.2 Land cover analysis panel

5. Next, navigate to **Landcover grid** (1) to select your land cover map.
6. Click **Save results as csv** (2), and select your preferred directory or folder to save your area statistics (3).
7. Under **Metric list**, click **Land cover**, and **Land cover** will appear under **Joblist** (5).
8. Click **OK** (5). The land cover area statistics will be calculated.
9. Next, open the csv file that you saved. This gives the area of each land cover class in square meters. You can easily convert square meters to square kilometers by diving 1,000,000 square meters.
10. Calculate the total area of all land cover classes, and fill this into the final column (Total).
11. Next, calculate the percent coverage (*Wi*) by dividing the strata total by the sum of all strata. Fill these values into the third row of the table.

	Bare areas	Built-up	Cropland	Grass/open areas	Water	Woodlands	**Total**
Area (km²)	76.4	317.4	972.7	1,149.1	24.9	313.3	2,853.8
W_i	2.68%	11.12%	34.08%	40.27%	0.87%	10.98%	

For example, 76.4 / 2,853.8 = 2.68%; 317.4 / 2,853.8 = 11.12%; 972.7/2,853.8 = 40.27%.

12. To determine the sample size for a stratified random sample, we will use the equation from Cochran (1977):

$$n \approx \left(\frac{\sum W_i S_i}{s\left(\widehat{P}\right)} \right)^2$$

where

W_i is the stratum weight;
S_i is the standard error for stratum i; estimated as $\sqrt{p_i(1 - p_i)}$;
p_i is the proportion of built-up area in stratum I;
$S(P)$ is the target standard error of the built-up area estimate.

12. We get the following information for determining the sample size assuming that one omission error of built-up area and a user's accuracy of 0.8 and a target standard error of the built-up area estimate of 0.5% (i.e., a confidence interval of 1%).

	Bare areas	Built-up	Cropland	Grass/open areas	Water	**Woodlands**
W_i	2.68%	11.12%	34.08%	40.27%	0.87%	10.98%
p_i	0.01	0.8	0.01	0.01	0	0.01
S_i	0.099	0.4	0.099	0.099	0	0.099
$S(P)$	0.005					
n ≈						

p_i for bare areas, cropland; grass/open areas and woodlands: 1 error of omission / 100 = 0.01.
S_i for bare areas, cropland; grass/open areas and woodlands: $\sqrt{0.01(1 - 0.01)}$= 0.099.
S_i for built-up areas: $\sqrt{0.8(1 - 0.8)}$= 0.4

14. Next, determine the sample based on the formula.

This in turn gives $n \approx \left(\frac{\sum W_i S_i}{s\left(\widehat{P}\right)} \right)^2 = \left(\frac{0.132}{0.005}\right)^2 = 676$ or in simplified form:

$$n \approx \left(\frac{0.0268 * 0.099 + 0.111 * 0.4 + 0.341 * 0.099 + 0.403 * 0.099 + 0.0087 * 0 + 0.1098 * 0.099}{0.005} \right)^2$$

$n \approx \left(\frac{0.132}{0.005}\right)^2 \approx (26)^2.$
$n \approx 676.$

15. The next step is to determine how to allocate these units to strata. Good practices recommend that 50–100 units are allocated to smaller land cover classes depending on the total sample size, while rest is proportionally allocated to the larger strata. Note that water and bare areas have fewer than 50 samples, while built-up areas and woodland areas have less than 100 samples. Therefore, we are going to increase built-up areas to 200 (since it is the one of the most important land cover classes) and bare areas and water to 50 samples. The final sample size will be 876.

	Bare areas	Built-up	Cropland	Grass/open areas	Water	Woodlands
proportional n	676 * 0.0268 = 18	676 * 0.111 = 75	676 * 0.341 = 230	676 * 0.403 = 272	676 * 0.0087 = 5	676 * 0.1098 = 74
final n_i	18 + 32 = **50**	75 + 125 = **200**	**230**	**272**	5 + 45 = **50**	**74**

6.2.2 Lab 2. Response Design

Objective

- To construct the confusion matrix and generate accuracy measures.

  Prerequisite—Install the AcATaMa plugin.

Procedure

In this lab, we are going to use AcATaMa, which is a QGIS plugin for performing accuracy assessment. The plugin is used to set up sampling design, assess the accuracy of thematic maps, and estimate areas of the map classes (Llano 2019). AcATaMa plugin is grouped into (i) Thematic, (ii) Sampling, (iii) Classification, and (iv) Accuracy Assessment panels.

The **Thematic panel** is used to set the thematic map, while the **Sampling design** defines how to select the sampled for the accuracy assessment (or any others uses). The plugin supports only *simple random sampling and stratified random sampling*. The **Classification** panel is used to select the sampling file, set up the **Grid setting** and **Classification** dialog. Last but not least, **Accuracy Assessment** panel computes the accuracy assessment measures. Following are the response design steps.

1. Go to *Layers > Add Raster Layer,* and load "Sent1_2_All_Variable_LC.img".
2. Right-click the land cover map in the layer panel, and click *Properties > Properties > Symbology* (Fig. 6.3). Set *Render Type* to *Paletted/Unique values,* and click *Classify.* Assign each land cover class an appropriate *Color* and *Label* (name).

Fig. 6.3 Layer properties. You can also save all layers style and config saving it in a QGIS project

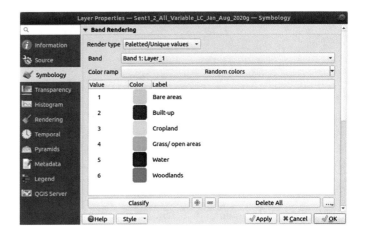

Fig. 6.4 AcATaMa plugin. *Note* The thematic map is the raster layer to which the accuracy assessment will be applied (e.g., a land cover map) and also is the base to generate the random sampling.

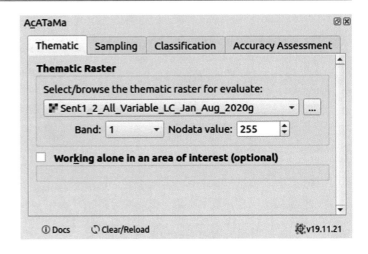

Fig. 6.5 Sampling methods

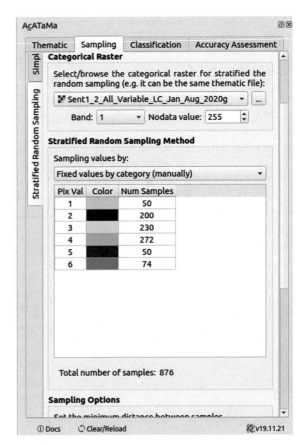

3. Click the *AcATaMa* plugin, and it appears (Fig. 6.4). Next, click **Thematic** panel, and under the Thematic Raster, load the "Sent1_2_All_Variable_LC.img"
4. Next, click the **Sampling** panel and click on the **Stratified Random Sampling Method** (Fig. 6.5). Under **Stratified Random Sampling Method** and *Sampling values by*, select the option *"Fixed values by category (manually)"*. Write the number of points for each stratum. Remember, you already determined the number of samples per land cover in the previous exercise.
5. You can set the minimum distance between the generated points (Fig. 6.6). For **With neighbors aggregation,** you can set the number of nearest neighbors pixels that belong to the same land cover class under **Number of neighbors.** This refers to the number of neighbors that AcATaMa evaluate to decide if a point can be included or not in the **Min neighbors with the same class**.

Fig. 6.6 Set up the options to restrain the allocation of the points

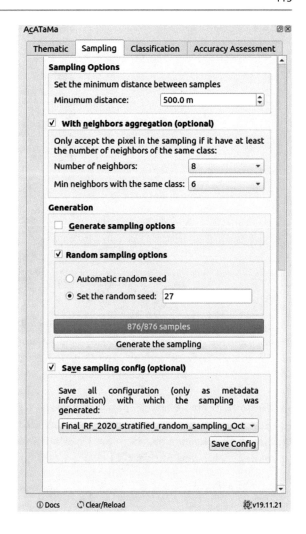

Fig. 6.7 Stratified random sampling points displayed on the land cover map

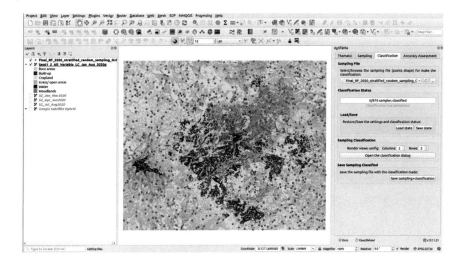

Fig. 6.8 Set thematic classes view12

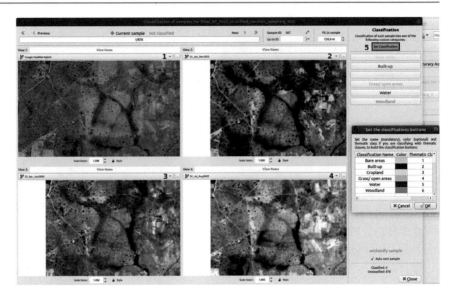

6. Next, input 27 *under the Set the random seed* (**Random sampling options**). The random seed value will generate reproducible sampling. Set the seed random number as an integer value.
7. Finally, click **Generate the sampling,** and click **Save Config**.
8. Next, navigate to the **Classification** panel. In the **Classification** panel, navigate to **Sampling File,** and select the sampling file. The stratified random sampling points are displayed on the land cover map (Fig. 6.7).
9. Go to **Sampling Classification,** and click on *Open the classification dialog*.
10. Select the images in views (steps 1–4), and set the **Fit to sample** (Fig. 6.8).
11. Set all classification buttons in **Set Classification** (step 5), and then, classify the samples using the *Set the classifications buttons* tab (steps 6) (Fig. 6.7).

Fig. 6.9 Save sample classification

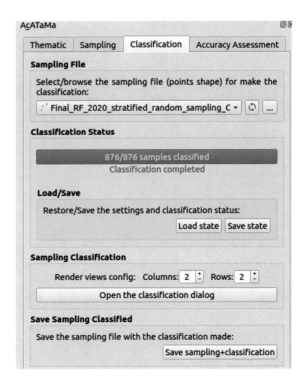

12. After configuring the **Set Classification**, click the land cover class that appears on Google Satellite Hybrid (step 1), S2_Jan_Mar2020 (step 2), S2_Apr_Jun2020 (step 3), and S2_Jul_Aug2020 (Step) as shown in Fig. 6.8.
13. Complete the classification, and save the classified samples under **Save Sampling Classified** (Fig. 6.9).
14. Navigate to the **Accuracy Assessment** panel, and load the classified sampling file under **Sampling File Classified** (Fig. 6.10).
15. Next, click *Open the accuracy assessment results* under the **Classification Accuracy Assessment**. This process creates the error or confusion matrix. The main diagonal of the error matrix highlights correct classifications, while the off-diagonal elements show omission and commission errors. Note that the total class area (km^2) and the column *Wi* (which shows the area proportion of each stratum in the map) are displayed (Fig. 6.11).

Fig. 6.10 Accuracy assessment panel

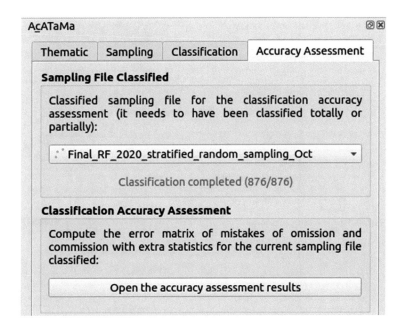

Fig. 6.11 Classification accuracy assessment results

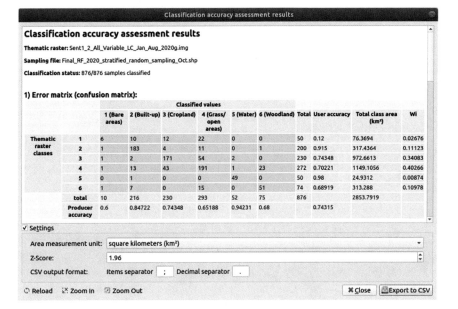

Classification accuracy assessment results

Thematic raster: Sent1_2_All_Variable_LC_Jan_Aug_2020g.img

Sampling file: Final_RF_2020_stratified_random_sampling_Oct.shp

Classification status: 876/876 samples classified

1) Error matrix (confusion matrix):

		Classified values						Total	User accuracy	Total class area (km²)	Wi
		1 (Bare areas)	2 (Built-up)	3 (Cropland)	4 (Grass/ open areas)	5 (Water)	6 (Woodland)				
Thematic raster classes	1	6	10	12	22	0	0	50	0.12	76.3694	0.02676
	2	1	183	4	11	0	1	200	0.915	317.4364	0.11123
	3	1	2	171	54	2	0	230	0.74348	972.6613	0.34083
	4	1	13	43	191	1	23	272	0.70221	1149.1056	0.40266
	5	0	1	0	0	49	0	50	0.98	24.9312	0.00874
	6	1	7	0	15	0	51	74	0.68919	313.288	0.10978
	total	10	216	230	293	52	75	876		2853.7919	
	Producer accuracy	0.6	0.84722	0.74348	0.65188	0.94231	0.68		0.74315		

☑ Settings

Area measurement unit:	square kilometers (km²)
Z-Score:	1.96
CSV output format:	Items separator ; Decimal separator .

↻ Reload   ⤢ Zoom In   ⊟ Zoom Out   ✖ Close   ⬛ Export to CSV

Fig. 6.12 Error matrix of estimated area proportion

Classification accuracy assessment results

2) Error matrix of estimated area proportion:

		1 (Bare areas)	2 (Built-up)	3 (Cropland)	4 (Grass/ open areas)	5 (Water)	6 (Woodland)	Wi
Thematic raster classes	1	0.00321	0.00535	0.00642	0.01177	-	-	0.02676
	2	0.00056	0.10178	0.00222	0.00612	-	0.00056	0.11123
	3	0.00148	0.00296	0.2534	0.08002	0.00296	-	0.34083
	4	0.00148	0.01924	0.06366	0.28275	0.00148	0.03405	0.40266
	5	-	0.00017	-	-	0.00856	-	0.00874
	6	0.00148	0.01038	-	0.02225	-	0.07566	0.10978
	total	0.00821	0.1399	0.3257	0.40292	0.01301	0.11026	

Header of the table above: **Classified values**

Fig. 6.13 Quadratic error matrix for estimated area proportion

Classification accuracy assessment results

3) Quadratic error matrix of estimated area proportion:

		1 (Bare areas)	2 (Built-up)	3 (Cropland)	4 (Grass/ open areas)	5 (Water)	6 (Woodland)
Thematic raster classes	1	0.0	0.0	0.0	0.0	-	-
	2	0.0	0.0	0.0	0.0	-	0.0
	3	0.0	0.0	0.0001	9e-05	0.0	-
	4	0.0	3e-05	8e-05	0.00013	0.0	5e-05
	5	-	0.0	-	-	0.0	-
	6	0.0	1e-05	-	3e-05	-	4e-05
	total	0.00291	0.00728	0.01343	0.0158	0.00257	0.00905

Header of the table above: **Classified values**

Fig. 6.14 Accuracy matrices

Classification accuracy assessment results

4) Accuracy matrices:

User's accuracy matrix of estimated area proportion:

		1 (Bare areas)	2 (Built-up)	3 (Cropland)	4 (Grass/ open areas)	5 (Water)	6 (Woodland)
Thematic raster classes	1	0.12	0.2	0.24	0.44	-	-
	2	0.005	0.915	0.02	0.055	-	0.005
	3	0.00435	0.0087	0.74348	0.23478	0.0087	-
	4	0.00368	0.04779	0.15809	0.70221	0.00368	0.08456
	5	-	0.02	-	-	0.98	-
	6	0.01351	0.09459	-	0.2027	-	0.68919

Header of the table above: **Classified values**

Producer's accuracy matrix of estimated area proportion:

		1 (Bare areas)	2 (Built-up)	3 (Cropland)	4 (Grass/ open areas)	5 (Water)	6 (Woodland)
Thematic raster classes	1	0.39099	0.03826	0.01972	0.02922	-	-
	2	0.06772	0.72752	0.00683	0.01518	-	0.00504
	3	0.18043	0.02119	0.77801	0.19861	0.22788	-
	4	0.18024	0.13756	0.19544	0.70176	0.11383	0.30879
	5	-	0.00125	-	-	0.65829	-
	6	0.18062	0.07423	-	0.05523	-	0.68616

Header of the table above: **Classified values**

Overall Accuracy:
0.72536

16. The absolute counts of the sample are converted into estimated area proportions (Olofsson et al. 2014) for systematic or stratified random sampling with the land cover map classes defined as the strata (Fig. 6.12).
17. The quadratic error matrix for estimated area proportion corresponds to the standard error estimated (Olofsson et al. 2014) (Fig. 6.13).
18. Figure 6.14 shows the classification accuracy assessment results.
19. Classes area adjusted table shows the uncertainty of the map area estimates (Fig. 6.15). Generally, the accuracy estimates adjust this estimate and also provide confidence intervals as estimates of uncertainty (Olofsson et al. 2014).
20. Finally, click on the **Export to CSV** button in order to save your accuracy assessment results.

Fig. 6.15 Class area adjusted results. *Note that by default the AcATaMa plugin calculates a 95% confidence interval (Z = 1,96). However, you can modify the z- score value according to the desired percent (Settings options in the report of results).*

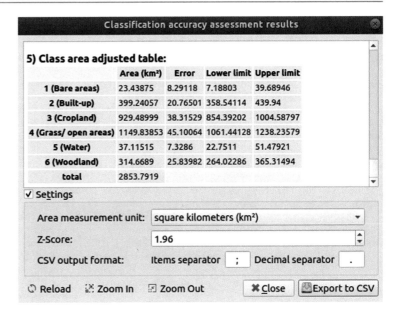

6.3 Summary

Rigorous (unbiased) classification accuracy assessment is an important indicator of the usefulness of a land cover map whether it has low or high accuracy. For example, an inaccurate map that has been rigorously assessed may provide highly valuable information, which can be used to improve land cover classification. Note that the rigorous accuracy assessment is calculated using area proportions not sample counts. This means that absolute counts of the sample are converted into estimated area proportions. Therefore, the rigorous accuracy assessment results are reported in actual area units (km^2 or hectares) and area proportions, which are more meaningful than mere pixel counts.

In this lab, we constructed the confusion matrix and generated accuracy measures. The error matrix (with the mapped areas of each map category) contains all the information needed to perform the analysis, which includes stratified estimation of area and confidence intervals. Here, we used the *AcATaMa plugin*. However, rigorous accuracy assessment can be done using other tools or software. For example, spreadsheet program can be used to provide the user with an understanding of the estimation procedure.

6.4 Additional Exercises

i. Perform rigorous accuracy assessment using the land cover map that you produced using support vector machine (SVM) classifier.
ii. Compare the accuracy assessment results, which were derived from final land cover maps that you produced using random forest and SVM classifiers.

References

Cochran WG (1977) Sampling techniques. Wiley, New York
Congalton R, Green K (2008) Assessing the accuracy of remotely sensed data: principles and practices, 2nd edn. CRC Press, Boca Raton
Congalton RG, Green K (1999) Assessing the accuracy of remotely sensed data: principles and practices. Lewis Publishers, Boca Rotan, Florida
Food and Agriculture Organization (2016) Map accuracy assessment and area estimation: a practical guide. National forest monitoring assessment working paper No.46/E
Foody GM (2002) Status of land cover classification accuracy assessment. Remote Sens Environ 80:185–201
Foody GM (2010) Assessing the accuracy of land cover change with imperfect ground reference data. Remote Sens Environ 114:2271–2285

Foody GM (2020) Explaining the unsuitability of the kappa coefficient in the assessment and comparison of the accuracy of thematic maps obtained by image classification. Remote Sens Environ 239:111–630

Jung M (2016) LecoS—a python plugin for automated landscape ecology analysis. Ecol. Inform 31:18–21

Llano XC (2019) AcATaMa—QGIS plugin for accuracy assessment of thematic maps, version XX.XX, https://plugins.qgis.org/plugins/AcATaMa/

Morales-Barquero L, Lyons MB, Phinn SR, Roelfsema CM (2019) Trends in remote sensing accuracy assessment approaches in the context of natural resources. Remote Sens 11(2305):1–16

Olofsson P, Foody GM, Stehman SV, Woodcock CE (2013) Making better use of accuracy data in land change studies: estimating accuracy and area and quantifying uncertainty using stratified estimation. Remote Sens Environ 129:122–131

Olofsson P, Herold M, Stehman SV, Woodcock CE, Wulder MA (2014) Good practices for estimating area and assessing accuracy of land change. Remote Sens Environ 148:42–57

Pontius RG Jr, Millones M (2011) Death to Kappa: Birth of quantity disagreement and allocation disagreement for accuracy assessment. Int J Remote Sens 32:4407–4429

Stehman SV (2009) Sampling designs for accuracy assessment of land cover. Int J Remote Sens 30(20):5243–5272

Stehman SV, Fonte CC, Foody GM (2018) Using volunteered geographic information (VGI) in design-based statistical inference for area estimation and accuracy assessment of land cover. Remote Sens Environ 212:47–59

Stehman SV, Foody GM (2019) Key issues in rigorous accuracy assessment of land cover products. Remote Sens Environ 231:1–23

Wulder MA, Franklin SE, White JC, Linke J, Magnussen S (2006) An accuracy assessment framework for large-area land cover classification products derived from medium-resolution satellite data. Int J Remote Sens 27(4):663–683

Appendix

A.1. Additional Learning Resources

There are several educational and training resources available to continue learning about GEE, R and machine learning.

Websites with machine learning, remote sensing and GIS exercises in GEE.

1. https://www.google.com/earth/outreach/learn/introduction-to-google-earth-engine/
2. https://developers.google.com/earth-engine/tutorials/tutorial_api_01
3. https://www.paulamoraga.com/tutorial-gee/
4. https://www.earthdatascience.org/tutorials/intro-google-earth-engine-ide/
5. https://www.geospatialecology.com/intro_rs_lab1/
6. https://ecology.colostate.edu/google-earth-engine/
7. https://geohackweek.github.io/GoogleEarthEngine/03-load-imagery/
8. https://courses.spatialthoughts.com/end-to-end-gee.html
9. https://github.com/giswqs/Awesome-GEE—by Qiusheng Wu

Websites with machine learning, remote sensing and GIS exercises in R.

1. Rtips. Revival 2014!
   http://pj.freefaculty.org/R/Rtips.html
2. Online R resources for Beginners
   2.1 http://www.introductoryr.co.uk/R_Resources_for_Beginners.html
3. About Quick-R
   https://www.statmethods.net/

Printed in the United States
by Baker & Taylor Publisher Services